KNOWLEDGE SPACES:
Theories, Empirical Research, and Applications

KNOWLEDGE SPACES: Theories, Empirical Research, and Applications

Edited by

Dietrich Albert
Karl-Franzens-Universität Graz
Graz, Austria

Josef Lukas
Martin-Luther-Universität Halle-Wittenberg
Halle/Saale, Germany

Psychology Press
Taylor & Francis Group
New York London

First published by Lawrence Erlbaum Associates, Inc.

This edition published 2012 by Psychology Press

Psychology Press
Taylor & Francis Group
711 Third Avenue,
New York, NY 10017

Psychology Press
Taylor & Francis Group
27 Church Road,
Hove, East Sussex BN3 2FA, UK

First issued in paperback 2014

Psychology Press is an imprint of the Taylor and Francis Group, an informa business

Cover design by Kathryn Houghtaling Lacey

Library of Congress Cataloging-in-Publication Data

Knowledge spaces : theories, empirical research, and applications /
[edited by] Dietrich Albert, Josef Lukas.
p. cm.
Includes bibliographical references and index.
ISBN 978-0-8058-2799-6 (hbk)
ISBN 978-1-138-00298-2 (pbk)

1. Natural language processing (Computer science) 2. Computational linguistics. 3. Artificial intelligence. 4. Knowledge, Theory of. I. Albert, Dietrich. II. Lukas, Josef.
QA76.9.N38K554 1999
006.3'5—dc21 99-11094
CIP

Contents

Contributors vii

Foreword ix

Preface xi

Acknowledgements xii

I: INTRODUCTION **1**

1. Knowledge Structures: What They Are and How They Can be Used in Cognitive Psychology, Test Theory, and the Design of Learning Environments. **3**
Josef Lukas and Dietrich Albert

II: THEORETICAL DEVELOPMENTS AND EMPIRICAL INVESTIGATIONS **13**

2. Component-based Knowledge Spaces in Problem Solving and Inductive Reasoning **15**
Dietrich Albert and Theo Held

3. Component-based Construction of Surmise Relations for Chess Problems **41**
Martin Schrepp, Theo Held, and Dietrich Albert

4. An Integrated Approach for Constructing, Coding, and Structuring a Body of Word Problems **67**
Theo Held

5. Modeling Knowledge as Competence and Performance **103**
Klaus Korossy

6. An Empirical Test of a Process Model for Letter Series Completion Problems **133**
Martin Schrepp

III: APPLICATIONS **155**

7. Organizing and Controlling Learning Processes Within Competence-Performance Structures **157**
Klaus Korossy

8. Structure and Design of an Intelligent Tutorial System Based on Skill Assignments **179**
Dietrich Albert and Martin Schrepp

9. Application of Doignon and Falmagne's Theory of Knowledge Structures to the Assessment of Motor Learning Processes **197**
Susanne Narciss

Author Index **221**

Subject Index **225**

CONTRIBUTORS

Dietrich Albert Institut für Psychologie, Karl-Franzens-Universität Graz, Universitätsplatz 2/III, A-8010 Graz, Austria.

Theo Held Institut für Psychologie, Martin-Luther-Universität Halle-Wittenberg, D-06099 Halle, Germany.

Klaus Korossy Bergheimer Str. 28, D-65115 Heidelberg, Germany.

Josef Lukas Institut für Psychologie, Martin-Luther-Universität Halle-Wittenberg, D-06099 Halle, Germany.

Susanne Narciss Institut für Pädagogische Psychologie und Entwicklungspsychologie, Technische Universität Dresden, Weberplatz 5, D-01062 Dresden, Germany.

Martin Schrepp Schwetzingerstraße 86, D-68766 Hockenheim, Germany.

FOREWORD

A scientific field in its infancy—if it is destined for maturity—requires loving care in the form of a steady diet of fresh ideas and results. This volume contributes to the development of knowledge spaces research by gathering an impressive quota of new applications, techniques and concepts.

The editors, Dietrich Albert and Josef Lukas, were among the first researchers in this field, to which they added important facets. One of their endeavours is especially noteworthy here. Knowledge space theory was initially formalized by Doignon and myself in lattice-theoretical terms with a relatively minimal concern for cognitive interpretations. This mathematical focus was deliberate on our part. For one thing, we were attempting to develop a formal language within which a priori vague concepts such as 'knowledge state' or 'learning path' could be defined and discussed without ambiguity. For another, we wanted this language to be applicable to widely diverse situations, a preoccupation which naturally led to favor abstraction. Early on, Albert and Lukas saw the need to equip this formal apparatus with a substantial content. In a series of papers, they strained to provide what they saw as the missing link between the mathematical aspects of the theory, and the underlying, mostly cognitive, content of the phenomena under study. Several chapters of this volume contain excellent illustrations of this particular strand of knowledge space research. I will comment on a representative example here.

In applying knowledge space theory to a particular empirical situation, the most critical stage certainly lies in the building of the structure, that is, in finding out what the feasible knowledge states are. Practical techniques have been developed by several researchers—e.g. Müller (1989), Koppen and Doignon (1990), Dowling (1993), Koppen (1993), see also Cosyn and Thiéry (to appear)[1]—, which essentially consist in a sophisticated questioning of experts in the field under study. Typically, the experts are not asked directly what the knowledge states are. Rather, their awareness of the feasible knowledge states is regarded as implicit, and tapped indirectly by questions such as: "Consider a student who regularly fails to solve Problem x. Would such a student be nevertheless capable of mastering Problem y"? From the collection of all the responses to questions like these (and some slightly more complex ones), a unique knowledge space can be con-

[1] Müller, C. E. (1989). A procedure for facilitating an expert's judgements on a set of rules. In E. E. Roskam (Ed.), *Mathematical Psychology in Progress* (pp. 157–170). New York: Springer.
Koppen, M., & Doignon, J.-P. (1990). How to build a knowledge space by querying an expert. *Journal of Mathematical Psychology, 34*, 311–331.
Dowling, C. E. (1993). On the irredundant generation of knowledge spaces. *Journal of Mathematical Psychology, 37*, 49–62.
Koppen, M. (1993). Extracting human expertise for constructing knowledge spaces: an algorithm. *Journal of Mathematical Psychology, 37*, 1–20.
Cosyn, E. & Thiéry, N. A practical procedure to build a knowledge structure. To appear in *Journal of Mathematical Psychology.*

structed, which is representative of that expert. This questioning is computerized, and quite tedious. It may take an expert several days to complete the questioning procedure for a domain of moderate size. In Albert and Lukas's team, a rather different tack is taken. Their choice of a knowledge structure is based on a systematic examination of the content of the problems. For example, they analyze the problems into components of difficulty, and formulate principles governing the combination of the components for any single problem. The knowledge structure is then derived from such an analysis. This technique is exemplified in Chapter 2 and 3 of this volume, which deal with inductive reasoning and chess problems. To be sure, as demonstrated by Cosyn and Thiéry in their paper cited above, a mixture of various techniques—some predicated on a content analysis, others relying on a questioning of experts, and last but not least, a statistical analysis of real assessment data based on some tentative knowledge structure—is likely to become, eventually, the recipe of choice for building a knowledge structure. (As anybody who has ever attempted such a task in a realistic situation would testify, this activity pertains, at least partly, to an art form.)

In general, this emphasis on content analysis as a guiding light to the knowledge structure, together with the various conceptual tools that they have introduced, may be the signature contribution of Albert, Lukas and their colleagues to the field of knowledge spaces. One of the consequences of their efforts is a much increased affinity between knowledge space theory and more traditional approaches of the cognitive sciences, certainly a valuable goal. Two other aspects of their work also contribute to this merging. They strive to connect their analysis to the main stream literature in the cognitive sciences, and they considerably extend the range of tasks analyzed in the language of knowledge space theory, from chess and word problems to letter series completion problems and—unexpectedly for this reader—motor skills.

Taken as a whole, the research reported in this volume much expand the wisdom in the field, both from a theoretical and an empirical standpoint, and is a welcome addition to the literature.

Jean-Claude Falmagne
March 12, 1999
Irvine, CA.

PREFACE

In the current age of information, education, knowledge, and learning continue to be of paramount importance to the individual and to society. Like other sciences, psychology, from its inception more than 100 years ago, has endeavored to understand and explain cognitive performances and the processes underlying cognitions in this field.

The most recent, and perhaps the most significant, approach in formal models for education and instruction was initiated by Doignon and Falmagne's development of the theory of knowledge spaces. This theory deals with knowledge-dependent performance and is focused on behavioral structures. It contributes to adaptive knowledge assessment, advanced computer tutoring systems, learning path models, and "intelligent" expert querying. It also provides the basics for inferring the structure of a knowledge system, deriving individual performance processes, assessing the individual's ability to perform cognitive subprocesses, and even validating process models empirically. By this, Doignon and Falmagne's framework constitutes the basis of the research program and the findings presented in the current volume.

The goal of this volume is to link observable, behavioral aspects to underlying concepts and to non-observable, psychological processes. The book presents the findings of programmatic research focused on the psychology of knowledge spaces. The research presented in this volume was carried out when the editors and authors were at the University of Heidelberg; the work was supported by the German Research Foundation's national program, which was focused on the psychology of knowledge.

Following an introduction to the field of knowledge structures (Chapter 1), various theoretical procedures pertaining to the derivation of knowledge spaces are presented together with relevant empirical findings. Chapters 2, 3, and 4 focus on the development of procedures for constructing knowledge spaces, taking into account the constraints imposed by the nature and ordering of problems in accord with various principles. Chapter 5 discusses how the competencies and skill structures of individuals needed to solve problems can be used to derive the structure of problems, deduce knowledge spaces, and assess the individual's performance states. In Chapter 6, knowledge spaces are derived from models of cognitive processing; these permit assessment of an individual's ability to perform cognitive subprocesses, and, most important, facilitate the empirical validation of process models.

From the very large range of possible problems, the editors and authors have selected a few examples for empirical investigation, specifically, several types of inductive reasoning and problem solving tasks. Learning processes provide the focus to the part of this volume that deals with applications. Selected topics include the organization of learning processes by competence structures (Chapter 7) and the design of an intelligent adaptive tutorial system based on skill structures

(Chapter 8). These chapters have theoretical, as well as practical, implications. In order to demonstrate the wide range of possible applications of the theoretical framework, the theory is also applied to the process of learning how to swim (Chapter 9). Of metatheoretical interest is the strong interdependence of psychological theory, empirical validation, and practical application demonstrated in this work.

ACKNOWLEDGEMENTS

We gratefully acknowledge the help of many others essential to the completion of this work. We are grateful to Jean-Claude Falmagne and Jean-Paul Doignon for bringing their epochal ideas to our attention early on. Of course, this book would not have been possible without the authors of the various articles and their careful work. Our thanks and appreciation go out to the German Research Foundation (Deutsche Forschungsgemeinschaft, DFG) for their financial support, and to Lawrence Erlbaum Associates for their friendly, professional services in publishing this book. Its typesetting in LaTeX has been particularly challenging. We thank Cord Hockemeyer (Graz) for his wonderful style file `LEAproc.cls` and Theo Held, Sebastian Krüger, Sabine Schmid and Karin Zelmer (Halle) for their efforts to get this book into shape. They put the various contributions together, reshaped paragraphs, tables, and figures, helped compiling the author and subject indices and finally succeeded in providing Lawrence Erlbaum Associates with the camera-ready copy in the required form. Last, but not least, we are grateful to Jean-Claude Falmagne, our spiritus rector, for writing the foreword to this book.

We believe that knowledge space theory is potentially the most influential psychological approach to progress in understanding knowledge, learning, and education. Whether our belief is validated will depend largely on our readers, specifically in their interest, motivation, and future contributions; their comments and suggestions for improving this book are more than welcome.

Graz/Halle, March 1999 Dietrich Albert and Josef Lukas

I

INTRODUCTION

1

Knowledge Structures: What They Are and How They Can be Used in Cognitive Psychology, Test Theory, and the Design of Learning Environments.

Josef Lukas
Martin-Luther-Universität Halle-Wittenberg

Dietrich Albert
Karl-Franzens-Universität Graz

In this first chapter of the book, we try to give a comprehensive introduction into the theory of knowledge structures. The general idea and the basic notation is outlined. In particular, we describe the background for our own approach to the formal theory. This approach may be characterized by three components: the *derivation* of a specific knowledge structure (by means of psychological task analysis, principles of test construction, individualized process modeling etc.), the *interpretation* of the theory within various knowledge domains, and its *application* in diagnostic and tutoring procedures. Thus, this chapter is an elementary introduction to the theory as well as an overview for the following contributions.

INTRODUCTION

In cognitive psychology we are interested in the question, how knowledge is represented internally. What is going on in our minds when we gather and store information about ourselves and about the world around us? How do we retrieve this information, modify it, and use it in everyday life?

In this context, the term *knowledge structure* is often used in a very unspecified way. Typically it means little more than the pure idea that knowledge has to be organized somehow. In this sense "structure" is a mere synonym for some kind of organization. Psychological theories concerning these organizing principles have concentrated on schemata, scripts, or on the properties of formal computer languages such as LISP or Prolog. Even in approaches that try to use mathematical definitions in cognitive psychology to some extent, the relationship of formal theoretical structure to observable behavior of individuals tends to be extremely vague.

The present book adopts a completely different view. It starts with observable behavior and is based on a formal, mathematical definition of what is meant by a knowledge structure. This deserves a careful explanation and motivation.

Mathematicians do have a clear and unequivocal understanding of the concepts of structure and structural properties. In the most general sense a structure is some set of entitities together with one (or more) relations and/or operations defined on it. Examples of elementary structures include ordinal structures (a set together with some order relation), topological structures (a set A together with a specified family of subsets, usually denoted the "open subsets" of A), or concatenation structures (a set A together with an operation $\circ : A \times A \longrightarrow A$). Obviously, interesting structures are comprised by various elementary ones. The real numbers $\boldsymbol{R}$ for example carry an ordinal structure ($>$), operations ($+, *$), and topological properties.

This is the background for the definition of a knowledge structure by Doignon and Falmagne (1985). A knowledge structure is a set of questions Q together with structuring relations defined on it. We are more specific in a moment. Right now, the only important point is to realize that the term knowledge structure is defined as a mathematical, formal structure, without any reference to its possible interpretation or to potential implementations in the mind. From a mathematical viewpoint the structure is pretty simple, some kind of so-called lattice and a priori it is far from obvious what the benefits are and how this approach may be useful in cognitive psychology.

The objective of the present chapter and the rest of the book is to attempt to give an answer to these questions already mentioned in Lukas and Albert (1993). We proceed as follows: In the next section, a first example is introduced. The following section provides some mathematical definitions. Only very basic concepts of order theory are needed. The remaining part of this chapter is devoted to consequences and applications of these definitions within cognitive psychology, test theory, knowledge assessment, and the design of learning environments. The various contributions in this book are natural consequences and detailed illustrations of the ideas developed in the present chapter. They describe theoretical extensions and report empirical investigations. Thus, in this introductory chapter, we try to provide the "red thread" for the contributions contained in this book.

TABLE 1
Example Problems for the Knowledge Domain "Elementary Calculus"

a) differentiate	$f(x) = x^2$
b) integrate	$f(x) = x^3 + x$
c) integrate	$f(x) = \frac{1}{1+x^2}$
d) differentiate	$f(x) = 3x^4 - 2x - 3$
e) differentiate	$f(x) = \frac{ax^3 - bx - c}{x^2}$
f) integrate	$f(x) = cosx$

A FIRST EXAMPLE

Let us start out with a first example of what is meant by a knowledge structure. Suppose you are a mathematics teacher and you want to get an idea of what your pupils know about elementary calculus, in particular their understanding of differentiation and integration of simple real-valued functions. The usual way to assess someone's knowledge is to put questions (in school as well as in psychological testing). The knowledge domain (differential calculus) is then defined as the set Q of all questions you might want to ask. Table 1 lists a few questions that might be contained in Q. Each question is answered either correctly or incorrectly. A structure is defined on Q as an order relation $\succeq$ with respect to the item difficulties. It might be reasonable to assume that question *e* is more difficult then *d*, question *c* is most difficult and so on. The ordinal structure of assumed item difficulties is depicted in Fig. 1.

SOME BASIC DEFINITIONS

To be more specific, it is natural to ask what kind of properties an order relation for item difficulties should have. A relation supposed to be interpreted as "... *is at least as difficult as...*" should obviously be a *quasi-order*, meaning that it should be *transitive* (if $a \succeq b$ and $b \succeq c$, then $a \succeq c$ for all objects a, b, c) and reflexive ($a \succeq a$ for all a). Quasi-orders can easily be visualized by means of so-called Hasse diagrams, where items with $a \succeq b$ are connected by directed lines and more difficult items are arranged above the easier ones. Note that some pairs

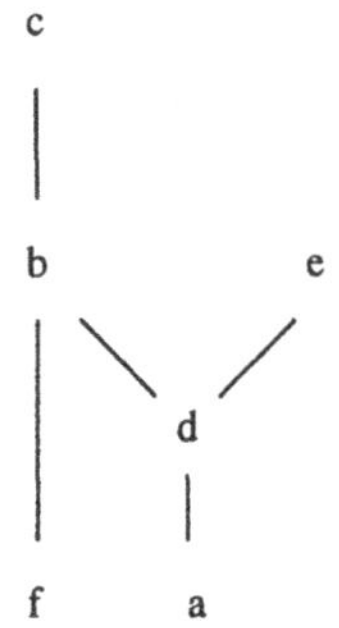

FIG. 1. Assumed difficulty structure for the items in Table 1 (see text for an explanation)

of items might not be connected at all, for example, item e and item f in Fig. 1. There are no a priori reasons why e should be more difficult than f (or vice versa) for all subjects.

A special case for a quasi-order is the well-know Guttman scale (Guttman, 1941), in which all items are linearly ordered along one single dimension. Disregarding any probabilistic errors and taking this relation as deterministic, the difficulty order has immediate consequences for the set of possible solution patterns. Let the quasi-order $\succeq$ be defined more concretely as

$$a \succeq b \quad \text{iff} \quad \begin{array}{l}\text{any subject who is able to solve } a \\ \text{should also be able to solve } b\end{array}$$

giving a clear operational definition for a being at least as difficult as b, and let $K \subseteq Q$ be the subset of solved problems for some particular subject. Then any K containing a should also contain b.

A subset of Q with this property will be called an *admissible solution pattern* or a solution pattern *compatible with* $\succeq$. Formally, the set of all admissible solution patterns is defined by

$$\mathcal{K} := \{K \subseteq Q \mid \text{if } a \in K \text{ and } a \succeq b, \text{ then } b \in K \text{ for all } a, b \in Q\}$$

Obviously, $\mathcal{K}$ as defined above is determined through the relation $\succeq$ and has some interesting algebraic properties. In particular, $\mathcal{K}$ is closed with respect to the union and intersection of sets. Conversely, any family of subsets closed under these two set operations defines uniquely a quasi-order. This result is due to Birkhoff (1937) and establishes a one-to-one correspondence between two formal structures, one being defined by an order relation and the other being a simple topological structure. The theory of knowledge spaces is based on a fundamental generalization of Birkhoff's theorem by Doignon and Falmagne (1985). We do not go into any mathematical details here, however. The mathematical concepts needed to comprehend the theoretical and empirical work described in this book are introduced and developed individually in every chapter. For anyone wishing to learn more about the mathematical background the recent book by Doignon and Falmagne (1998) provides an excellent introduction as well as a complete and extensive treatment of the formal concepts underlying the theory of knowledge spaces. Right here it suffices to note that a knowledge structure is characterized either as some order structure or, alternatively, a complete list $\mathcal{K}$ of all admissible solution patterns. The order structure is usually a quasi-order $\succeq$ (called a *surmise relation*) or, more generally, a mapping σ called a *surmise system* that associates to each element a of Q a nonempty collection of subsets of Q.

In this context, the present book is concerned with two major topics, namely:

1. How is a knowledge structure determined for a given knowledge domain? Various techniques are described, providing a particular structure on a set of problems including: *task analysis* revealing relevant problem components (formal descriptions of problem characteristics), formulation of *skills* (cognitive elements) necessary and sufficient for a solution of the problem under scrutiny, and complete *cognitive theories* allowing the derivation of item difficulty relationships.
2. The second major topic investigates application perspectives, namely, empirical evaluation of cognitive theories, algorithms for diagnostic procedures, and learning environments.

DETERMINATION OF A KNOWLEDGE STRUCTURE

Application of the theory of knowledge structures requires a set of problems to ask, and explicit assumptions about the knowledge structure, for example, the surmise relation defined on this set. How can this information be determined for a given knowledge domain?

A couple of different approaches have been suggested for this purpose. Van Leeuwe (1974), for example, describes a method called *item tree analysis*, in which a direct estimation of the surmise relation is based on an inspection of empirical contingency tables for all possible pairs of problems. Similar suggestions have been made in more general approaches to binary data analysis by Airasian and Bart (1973), van Buggenhaut and Degreef (1987) and Flament (1976). Its applicability in knowledge assessment, however, is limited. The hypothetical problem structure resulting from binary data analysis strongly depends on the choice of appropriate parameter values (e. g. frequency thresholds) and on the sample of subjects. For a detailled discussion comparing item tree analysis with alternative approaches in knowledge assessment see Held, Schrepp, and Fries (1995). A slightly different method is based simply on counting the frequencies of all possible patterns in large empirical data sets. If the observed frequency for a particular pattern exceeds a specified limit, then this pattern is taken to be admissible. This method is demonstrated for example by Falmagne (1989), Müller and Regenwetter (1991), and Villano (1991). Obviously, it is feasible only in situations with few problems and large data sets.

A second method of establishing $\mathcal{K}$ is to ask experts in or teachers of the knowledge domain about the difficulty structure of the problems. Koppen and Doignon (1990) and Müller (1989) developed algorithms for computerized query procedures. A series of refinements and improvements is reported by Kambouri, Koppen, Villano, and Falmagne (1994), Koppen (1993), Dowling (1993), and others. Still, this direct method of determining a knowledge structure is tedious and time consuming, even for moderately sized problem sets.

In our own group, we developed an alternative approach to determine $\mathcal{K}$. Rather than asking experts about their (implicit) assumptions concerning the structural relationships between the test items, we endeavored to formulate *explicitly* the argumentation for these relationships. A good example is provided by Albert and Held (Chapter 2). For a set of 12 number series completion problems, so-called *components* are extracted that are supposed to account for the relative problem difficulties. The knowledge structure is then derived from explicitly formulated principles about component combinations contained in each problem. Please note, in particular, that in this study (as well as in all other chapters in this volume) the assumptions are tested empirically and thus are subject to a rigorous empirical validation.

A similar approach is suggested in Schrepp, Held, and Albert (Chapter 3). The components considered here are tactical elements in chess (so-called motives) whose combinations allow the prediction of the solvability of simple chess problems. Obviously, these components are extremely domain-specific and experts in this field have to be consulted. Unlike the above-mentioned query procedures, we do not ask for problem orderings directly. Only sophisticated and problem-related questions are asked, using domain-specific vocabulary (e. g.,"Which *motives* are contained in this problem") rather than general questions about problem difficulties. The questions to the experts are more specific, they are usually perceived as being more reasonable and—above all—they are restricted in number.

A similar idea uses assumptions about *skills* necessary and sufficient to solve questions or problems. Korossy (Chapter 5) develops an extensive formal description of skill-structures (called competence), solution structures (called performance) and the relationship between both domains (called diagnostics). Several algebraic properties of diagnostics are derived and illustrated with empirical data on problems in elementary school geometry.

Narciss (Chapter 9) pursuits a similar path. Besides surmise relations Narciss uses *incompatibility structures* suggested by Lukas (1997) as a second principle. Its particular appeal and originality comes from the specific domain of application in sports. Highly trained breast stroke swimmers have to judge diagrams of arm and leg positions according to the correctness of their coordination. The results are compared to "ideal structures" derived from a biomechanical analysis, the actual swimming performance of the subjects and the effects of mental training on both, the theoretical knowledge about optimal coordination and the performance in competition situations.

A natural "next step" in this direction is an attempt to formulate a complete cognitive theory about the underlying solution process. A classical example has been presented by Spada (1976), who tried to derive difficulty parameters for items of a specific intelligence test from assumptions about an algorithmic treatment of the tasks performed by the subjects. This approach has been confined to one-dimensional item characteristics, whereas the following examples are based on knowledge space theory and thus allow models with very general ordinal structures.

Schrepp (Chapter 6), for example, refers to an algorithm suggested by Simon and Kotovsky (1963) for the solution of letter series completion problems. The knowledge structure can be derived from the process parameters in a straightforward manner and is shown to be a valid predictor of solution data in an empirical investigation.

Held (Chapter 4) uses problems from elementary stochastic courses and studies the influence of textual properies of word problems on the solution performance. His knowledge structures are based on the theory of Kintsch and van Dijk (1978) on text comprehension, in particular on "coherence" and "cohesion" of the problem's formulation.

Both studies show an interesting and far-reaching special application of knowledge structure theory: It can be used to test the validity of a cognitive psychological theory empirically by considering individual differences. The general procedure is comprised of three steps:

- Formulate a model for the solution process explicitely enough, such that assumptions for the knowledge structure can be derived.
- Derive the knowledge structure (set of admissible solution patterns).
- Check observed solution patterns against the set of admissible combinations defined by the knowledge structure.

APPLICATION OF KNOWLEDGE STRUCTURES

A first application has been shown in the previous paragraph. Knowledge structures are used to validate theoretical assumptions or hypotheses about cognitive processes by comparing empirical solution patterns with the admissible patterns derived from the assumptions. Knowledge space theory provides a formal framework for this test and forces a clear and operational formulation of the theory.

A second class of applications is concerned with test theory and knowledge assessment. If, for a given knowledge domain, the structural properties are known, then this information can be used for the design of a tailored testing procedure.

The general idea is to choose the next problem in an assessment procedure depending on the previous responses of the subject. The *most informative next item* has to be found, which is the item that reduces the uncertainty about this particular subject in this particular situation maximally. For this purpose, the knowledge structure, which is actually some sort of item difficulty structure rather than an individual knowledge organization, can be used. Falmagne and Doignon (1988b) described a Markovian procedure for knowledge assessment that can be implemented in any diagnostic system with a given knowledge structure. A similar stochastic algorithm is presented in Falmagne and Doignon (1988a). In this volume, two chapters are concerned with applications in test theory. Korossy (Chapter 5) suggests a very general, formal definition of the term *diagnostic* as a mapping between performance and competence structures and discusses the consequences for assessment procedures. Thus, tailored testing methods can also be used for assessing individual knowledge organization.

The results reported in Held (Chapter 4) are also aiming at specific application aspects. His method of deriving the position of every single item in a knowledge structure from its components provides the possibility of online item constructions during testing. The next most informative item to be presented is *generated* by an appropriate component combination individually.

Knowledge assessment frequently occurs within learning environments. Intelligent tutoring systems, for example, must usually have diagnostic components. This refers to the third class of applications intended by knowledge space theory. One of the key papers here is the formulation of *stochastic learning paths* in Falmagne (1989). Two chapters in this volume are dealing with learning processes within knowledge structures. Korossy (Chapter 7) describes knowledge acquisition in the context of his competence-performance structures introduced in Chapter 5. Albert and Schrepp (Chapter 8), on the other hand, investigate the consequences of skill assignments for intelligent tutoring systems.

To summarize, the theory of knowledge spaces, originated by Doignon and Falmagne (1985) and extended by our group as described in this book, results in different psychological principles for deriving empirically testable knowledge structures, including component-based task analysis, structured skill assignments, and the individualizing of process models. An empirical validation of the obtained structures validates at the same time the underlying psychological theories, because the relationship between empirical structures and the theoretical structures of cognitive demands, cognitive skills, and cognitive processes is described precisely and explicitly. Thus, a new methodology for testing psychological hypothesis and theories by taking individual differences into account is provided by extending the formalism of knowledge space theory.

In addition, the different psychological approaches for deriving empirically testable knowledge structures are guiding principles in the field of test construction. The obtained tests are supposed to assess an individual's performance, skills or cognitive functioning differentially and more efficiently than classical procedures. They can even be used for the assessment of misconceptions and mental states and provide guidelines for a description of knowledge acquisition as well as for the definition and the design of intelligent tutoring systems.

ACKNOWLEDGEMENTS

The research reported in this paper was supported by Grant Lu 385/1 of the Deutsche Forschungsgemeinschaft to J. Lukas and D. Albert at the University of Heidelberg.

REFERENCES

Airasian, P. W., & Bart, W. M. (1973). Ordering theory: A new and useful measurement model. *Educational Technology, May*, 56–60.

Birkhoff, G. (1937). Rings of sets. *Duke Mathematical Journal, 3*, 443–454.

Doignon, J.-P., & Falmagne, J.-C. (1985). Spaces for the assessment of knowledge. *International Journal of Man-Machine Studies, 23*, 175–196.

Doignon, J.-P., & Falmagne, J.-C. (1998). *Knowledge spaces*. Berlin: Springer.

Dowling, C. E. (1993). On the irredundant generation of knowledge spaces. *Journal of Mathematical Psychology, 37*, 49–62.

Falmagne, J.-C. (1989). A latent trait theory via a stochastic learning theory for a knowledge space. *Psychometrica, 54*, 283–303.

Falmagne, J.-C., & Doignon, J.-P. (1988a). A class of stochastic procedures for the assessment of knowledge. *British Journal of Mathematical and Statistical Psychology, 41*, 1–23.

Falmagne, J.-C., & Doignon, J.-P. (1988b). A markovian procedure for assessing the state of a system. *Journal of Mathematical Psychology, 32*, 232–258.

Falmagne, J.-C., Koppen, M., Villano, M., Doignon, J.-P., & Johannesen, L. (1990). Introduction to knowledge spaces: how to build, test, and search them. *Psychological Review, 97*, 201–224.

Flament, C. (1976). *L'analyse booléenne de questionnaire*. Paris: Mouton.

Guttman, L.A. (1941). The quantification of a class of attributes: A theory and method of scale construction. In P. Horst (Ed.), *The prediction of personal adjustment* (pp. 319–348). New York: Social Science Research Council.

Held, T., Schrepp, M., & Fries, S. (1995). Methoden zur Bestimmung von Wissensstrukturen — Eine Vergleichsstudie [Methods for the determination of knowledge structures — a comparison study]. *Zeitschrift für Experimentelle Psychologie, 42*, 205–236.

Kambouri, M., Koppen, M., Villano, M., & Falmagne, J.-C. (1994). Knowledge assessment: tapping human expertise by the QUERY routine. *International Journal of Human-Computer-Studies, 40*, 153–184.

Kintsch, W., & van Dijk, T. A. (1978). Toward a model of text comprehension and production. *Psychological Review, 86*, 363–394.

Koppen, M. (1993). Extracting human expertise for constructing knowledge spaces: an algorithm. *Journal of Mathematical Psychology, 37*, 1–20.

Koppen, M., & Doignon, J.-P. (1990). How to build a knowledge space by querying an expert. *Journal of Mathematical Psychology, 34*, 311–331.

Lukas, J. (1997). Modellierung von Fehlkonzepten in einer algebraischen Wissensstruktur [Modeling misconceptions in an algebraic knowledge structure]. *Kognitionswissenschaft, 4*, 196–204.

Lukas, J., & Albert, D. (1993). Knowledge assessment based on skill assignment and psychological task analysis. In G. Strube & K. F. Wender (Eds.), *The Cognitive Psychology of Knowledge*, Vol. 101 of *Advances in Psychology* (pp. 139–160). Amsterdam: Elsevier.

Müller, C. E. (1989). A procedure for facilitating an expert's judgements on a set of rules. In E. E. Roskam (Ed.), *Mathematical Psychology in Progress* (pp. 157–170). New York: Springer.

Müller, H. R., & Regenwetter, M. (1991). *A Procedure for facilitating an expert's judgments on a set of rules.* Official final report of the research project: "Systèmes dynamiques simulés par ordinateur", MEN/IPE/87/009, Ministère de l'Education Nationale, Luxembourg.

Simon, H. A., & Kotovsky, K. (1963). Human acquisition of concepts for sequential pattern. *Psychological Review, 70*, 534–546.

Spada, H. (1976). *Modelle des Denkens und Lernens.*[Models of thinking and learning]. Bern: Hans Huber.

Van Buggenhaut, J., & Degreef, E. (1987). On dichotomisation methods in boolean analysis of questionaires. In E. E. Roskam & R. Suck (Eds.), *Progress in mathematical psychology* (pp. 447–453). Amsterdam: Elsevier.

Van Leeuwe, J. F. J. (1974). Item tree analysis. *Nederlands Tijdschrift voor de Psychologie, 29*, 475–484.

Villano, M. (1991). *Computerized knowledge assessment: Building the knowledge structure and calibrating the assessment routine*, Doctoral Dissertation, New York University, New York.

II

THEORETICAL DEVELOPMENTS AND EMPIRICAL INVESTIGATIONS

2

Component-based Knowledge Spaces in Problem Solving and Inductive Reasoning

Dietrich Albert
Karl-Franzens-Universität Graz

Theo Held
Martin-Luther-Universität Halle-Wittenberg

Two principles for component-based problem construction and ordering are presented and applied to different knowledge domains. The first principle is called "set inclusion." It is used for constructing and ordering a set of chess problems. The second principle is "componentwise ordering of product sets." We apply it in the domain of inductive reasoning (i. e. number-series completion). While set inclusion did not prove to be adequate for an application in the considered domain, the componentwise ordering of product sets led to remarkably good results in two empirical investigations.

INTRODUCTION

Procedures that are to test a subject's knowledge concerning a specific domain obviously require—in addition to other prerequisites—a set of problems. The answers to these problems may serve as a basis for a hypothesis about the subject's actual knowledge. A teacher might assume that a student possesses all of the knowledge necessary to solve the problems. There are at least two different methods of questioning:

- All available problems are presented and the set of problems which have been solved correctly is assumed to represent the student's knowledge concerning the investigated domain. This method seems to be rather uneconomical, particularly if the set of problems is quite large.
- The problems that are presented are selected adaptively from a problem set. If a teacher presents a problem which is solved correctly by a student, the next problem will probably be more difficult because the teacher will suppose that the student is capable of solving all easier problems.

Certainly, the second method of knowledge assessment requires an a-priori hypothesis about a *structure* on the problem set. Such a hypothesis may, for example, be: "If a student succeeds in multiplying two fractions, she or he will also be able to multiply two natural numbers." The manner in which a teacher will conduct an assessment procedure depends largely on his or her own experience and knowledge. This experience and knowledge is implicitly used for structuring a knowledge domain. We investigate these hypotheses of a domain's structure in a formal way. First, we take a look at various types of relations that may be defined on a set of problems. This overview serves as a prerequisite for a short introduction to the theory of knowledge spaces put forward by Doignon and Falmagne (1985). Then we will focus on the question of how a relation on a set of problems can be established by using principles for systematical problem construction and discuss two empirical examples.

First, we give some examples of relations on sets of problems. For this purpose, we must introduce a few basic concepts of ordering theory, for example, how can statements like "problem x is more difficult than problem y" or "problem x is at least as difficult as problem y" be denoted?

Most of our examples will involve special cases of quasi-orders (reflexive and transitive). These are partial orders (reflexive, transitive, and antisymmetric), weak orders (transitive and connected), linear orders (connected, antisymmetric, and transitive), and antichains (reflexive, transitive, symmetric, and antisymmetric). In brackets, the defining properties are given from which other properties can be logically derived (e. g., reflexivity of linear and weak orders follows from connectedness). Sometimes irreflexive orders are also used. For a general introduction to ordering theory we recommend of Davey and Priestley (1990).

EXAMPLE 2.1 On a set $Q = \{w, x, y, z\}$ of problems a linear order $\{(w, w)$, (x, x), (y, y), (z, z), (x, w) (y, x), (z, y), (z, x), (z, w), $(y, w)\}$ is defined. First, let us look at the Hasse diagram in Fig. 1(a). We can see that this order is a special case of a quasi-order because every problem is comparable to all other problems. Problem w, for example, is supposed to be more difficult than problems x, y, and z. This type of problem ordering is known in psychology as a *Guttman scale* (Guttman, 1947, 1950). □

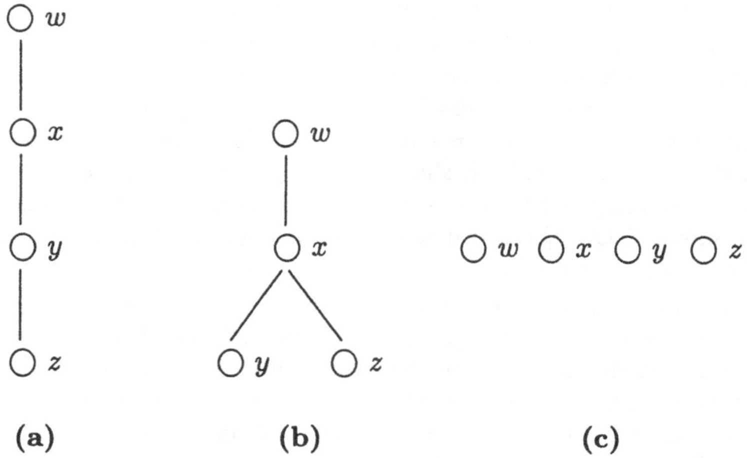

FIG. 1. Hasse diagrams for Examples 2.1, 2.2 and 2.3.

EXAMPLE 2.2 $Q = \{w, x, y, z\}$ is a set of problems with a quasi-order $\{(w, w), (x, x), (y, y), (z, z), (x, w), (y, x), (z, x), (y, w), (z, w)\}$ defined on this set. This order is shown in Fig. 1(b). For our set of questions, this means that problem x should be more difficult than problem y and problem z. It is assumed that problems y and z cannot be compared. □

EXAMPLE 2.3 On a set $Q = \{w, x, y, z\}$ of problems an antichain order $\{(w, w), (x, x), (y, y), (z, z)\}$ is defined. The Hasse diagram in Fig. 1(c) shows that the problems of Q are not connected. Nevertheless, this relation is also a special case of a quasi-order. The postulation of an antichain order may be adequate for sets of heterogeneous and completely incomparable problems. However, it is clear that such a set cannot be used for an economical adaptive questioning procedure because from a subject's answers, no conclusions on other answers can be drawn. □

An important topic in problem ordering is the interpretation of the binary relation that is defined on the problem set. Up until now, we have mentioned only a few unspecified differences in difficulty. In addition to further theoretical considerations, an interpretation will be introduced in the following section.

PROBLEM STRUCTURES AND KNOWLEDGE SPACES

The theory of knowledge spaces (Doignon & Falmagne, 1985, 1998) shows how the structure of problems can be represented in a formal way (for an introduction, see Falmagne, Koppen, Johannesen, Villano & Doignon, 1990).

Let $Q = \{w, x, y, z\}$ be a set of problems that is used for an examination. For some of these problems, a statement such as *"if a student is able to solve a specific problem in Q, he or she will also be able to solve other problems belonging to Q"* may be plausible. This can be formalized in terms of a binary relation $\preceq$. The expression $(y, x) \in \preceq$, which is abbreviated by $y \preceq x$ is interpreted as follows: Given a correct response to problem x, we *surmise* a correct answer to problem y. The relation $\preceq \subseteq Q \times Q$ is called *surmise relation*. It is assumed that the surmise relation is a quasi-order on Q.

A surmise relation can be depicted as a Hasse diagram. The diagrams shown in Fig. 1 can be interpreted as hypothetical orders on the problem set Q. According to the order shown in Fig. 1(b), we assume that each of the students capable of solving problem x will also be able to solve problem y *and* problem z. Based on this assumption, we can collect all subsets of Q that agree with the surmise relation. These subsets are called *knowledge states*.

DEFINITION 2.1 (see Falmagne et al., 1990) Let Q be a set of problems. $K \subseteq Q$ is a *state* $\Leftrightarrow$ $(\forall q, t \in Q, q \preceq t \wedge t \in K \Rightarrow q \in K)$. □

The family of all possible states with respect to a set of problems is a *knowledge structure*. For Example 2.2, we obtain the structure $\mathcal{F}$:

$$\mathcal{F} = \{\emptyset, \{y\}, \{z\}, \{y, z\}, \{x, y, z\}, \{w, x, y, z\}\}.$$

This knowledge structure contains all subsets of Q that are expected to occur as results of diagnostic procedures. The purpose of such procedures is to assign subjects to one of these states without presenting all problems in Q (see Falmagne et al., 1990). $\mathcal{F}$ is closed under union and intersection. A knowledge structure with these properties is called a *quasi-ordinal knowledge space*. A one-to-one correspondence between transitive and reflexive orders and families of knowledge states that are closed under union and intersection is established by a theorem by Birkhoff (1937). The restriction of closure under union and intersection is somewhat unrealistic for many knowledge domains. Therefore, Doignon and Falmagne introduced, as a generalization of quasi-ordinal knowledge spaces, the concept of *knowledge spaces*. Knowledge spaces are families of states that are closed under union, but do not have to be closed under intersection. Hence, every quasi-ordinal knowledge space is also a knowledge space. Doignon and Falmagne showed that there is a one-to-one correspondence between knowledge spaces and the so-called *surmise systems*. This will not be discussed in detail here. Our further considerations will deal solely with quasi-ordinal knowledge spaces.

Quasi-ordinal knowledge spaces can also be derived from "special cases" of quasi-ordered problem sets such as sets with an antichain order or linearly ordered sets.

"COMPONENT-BASED" ESTABLISHMENT OF SURMISE RELATIONS

We will expand our considerations about problems with a topic we call *problem component* or simply *component*. One way to facilitate problem comparison is by *systematical problem construction*. Construction principles are applied on well-defined sets of problem components. Furthermore, by means of their associated component structures, we can both provide a precise description of problems and the class of possible problem variations. Certainly, components have to be equipped with properties that are prerequisites for a successful combination.

Before we introduce two construction principles, a short sketch of the concept that we call a problem component should be drawn. As an example, let us imagine we are asked to solve an algebraic problem, for example, the multiplication of two fractions. Although this is a simple task, we will not be able to give the solution if we do not know some basics of algebra. Some of these basics may be "multiplication of natural numbers", "division of natural numbers", and "rules for the multiplication of fractions."

These items can be seen as *cognitive demands* on a subject confronted with the problem. If the subject does not have the knowledge at his or her disposal which is "demanded" or if the subject is not able to apply this knowledge, it is supposed that the answer to the problem will be incorrect—assuming the guessing probability is equal to zero.

The principle of set inclusion

We now take into account the representation of problems as sets of components. The following examples give a first idea of how problems can be constructed from components.

EXAMPLE 2.4 Let $C = \{a, b, c\}$ be a set of problem components. We assume that an antichain order is defined on C. Let us identify problems with subsets of C. With respect to the antichain order defined on the components, we assume that no dependencies between components exist. Hence, every subset of C can be identified with a potential problem and thus, with an element of a problem set Q (in this case subsets of C denote problems!):

$$Q = \{\emptyset, \{a\}, \{b\}, \{c\}, \{a,b\}, \{a,c\}, \{b,c\}, \{a,b,c\}\}.$$

The combination of the components a, b, and c has led to seven problems. The "empty problem" $\emptyset$ is left out because it cannot be shown. We assume that a problem is more difficult than another problem if it is characterized by all components of the other problem and by at least one more component. According to this assumption, we can state a hypothetical order as shown in Fig. 2(a). The next

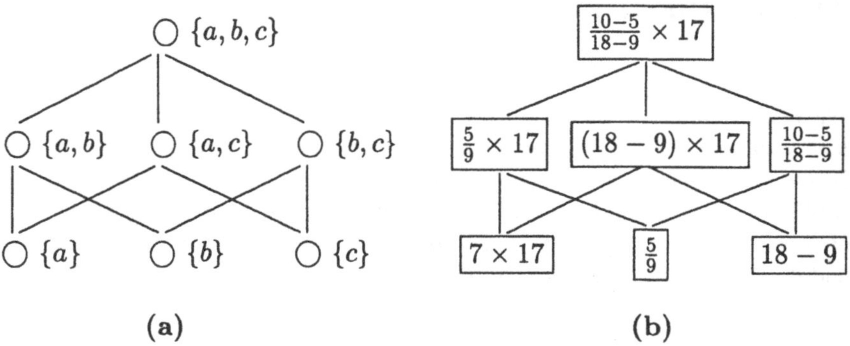

FIG. 2. Problem structure for the questions in Q and an example concerning calculation problems.

step is the application of the construction principle to concrete problem components and the ordering principle on concrete problems. For example, we define for a, b, c: $a \mathrel{\widehat{=}}$ multiplication of numbers, $b \mathrel{\widehat{=}}$ division of numbers, $c \mathrel{\widehat{=}}$ subtraction of numbers. Hence, $\{a, b, c\}$ is a problem that contains multiplication, division and subtraction, for example

$$\frac{10-5}{18-9} \times 17.$$

The hypothetical order for a set of problems is shown in Fig. 2(b). □

We have to stress that a surmise relation was established by constructing and ordering the problems in this way. Unconnectedness, however, is not a necessary property of the component set. As we will show, this method is also applicable to linearly ordered and quasi-ordered component sets. Examples 2.5 and 2.6 give an idea of this method.

EXAMPLE 2.5 We have a set $C = \{a, b, c\}$ of linearly ordered problem components. Fig. 3 shows, on the left, a possible Hasse diagram for the component structure and, on the right, the structure of the resulting problems (components are marked by triangles). Taking into account that the elements of C are linearly ordered, only three problems can be constructed. This linear order may, for example, be induced by constraints on combining the components. Such constraints can exist for sets of non-independent components. In our example, a may be a problem component that also contains b and c in some way. Therefore, if one part of a problem is associated with a, then b and c are automatically involved. To illustrate this we assume that component a corresponds to the addition of natural numbers within the hundreds, for example $619 + 347$. Furthermore we assume that b corresponds to the addition of the numbers between one and ten. We see that b is also necessarily an element of a. Thus, b is an element of problems containing a. □

EXAMPLE 2.6 We assume that a quasi-order is defined on a set $C = \{a, b, c\}$ of problem components. Fig. 4 shows one of the possible Hasse diagrams (left) and the corresponding problem structure (right). From this quasi-ordered problem set, that corresponds to the order shown in Example 2.2, single-component problems consisting either of b or c can be constructed. We may suppose that b and c are thematically independent, but are both involved in a in some way. □

After this brief introduction to one possible method of constructing and ordering problems by means of problem components, we state these principles formally:

DEFINITION 2.2 Let C be a set of components and $\preceq$ a quasi-order on C. The *component space* $\mathcal{F}_C$ is the family of all subsets T of C for which

$$x \in T, y \preceq x \Rightarrow y \in T$$

holds. □

Given a component space $\mathcal{F}_C$, according to each element T of $\mathcal{F}_C$ a problem q_T is formulated. A *surmise relation* R on the problem set $Q = \{q_T \mid T \in \mathcal{F}_C\}$ is defined by the following condition:

$$q_T \, R \, q_{T'} :\Leftrightarrow T \subseteq T'.$$

This means that the problems are identified with the elements of $\mathcal{F}_C$, while the relation R is identified with $\subseteq$.

It is easy to verify that R is a transitive relation: Let M, M', M'' be sets with $M \subseteq M'$ and $M' \subseteq M''$. M' which contains M is a subset of M'', thus $M \subseteq M''$, which means that '$\subseteq$' is transitive.

This ordering principle of set inclusion is based on the plausible assumption that a subject succeeding in the solution of a given problem will also be able to solve all the subproblems of this problem. We have to note here that a reversed

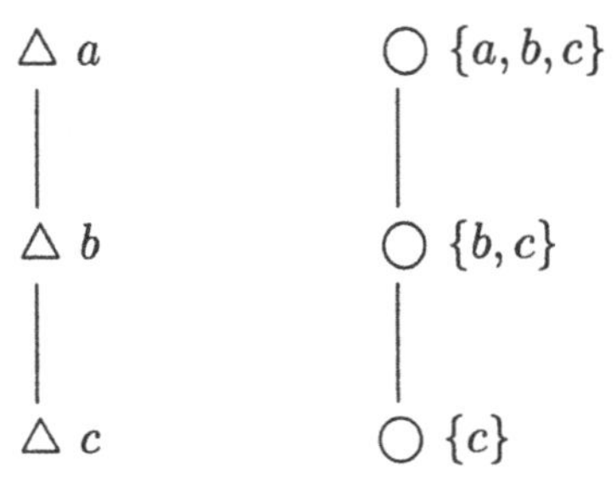

FIG. 3. Component structure and problem structure for Example 2.5.

statement such as "if someone is able to solve the subproblems, she or he will also be able to solve the superset problem" is not expected to hold true. The combination of problem components may lead to some additional difficulties that might not appear within the single components.

As an ordering method, set inclusion can be applied to very different theoretical approaches in the field of knowledge assessment. For examples, we refer to the investigations of Korossy (1993, 1997).

The principle of componentwise ordering of product sets

Up until now, we have focused only on single sets of components that were characterized by an order that was defined on the component set. In this section, we turn our attention to the construction of problems that consist of components with variable attributes. Here, every problem is equipped with the same number of components. New problems are constructed by varying the attributes of the components. The order of the problems will be derived from relations that are defined on the set of attributes. The following example gives an idea of this method.

EXAMPLE 2.7 Let $A = \{a_1, a_2, a_3\}$ and $B = \{b_1, b_2\}$ be problem components; a_1, a_2, a_3 and b_1, b_2 are the attributes of these components. On both sets A and B, a linear order is defined (see the left side of Fig. 5; attributes are marked by black triangles).

Suppose we want to construct simple algebra problems. One component may be the set of numbers that is used within a calculation, the other component is characterized by the operations that are to be applied on the set of numbers. We define: $a_1 \mathrel{\widehat{=}}$ use of real numbers, $a_2 \mathrel{\widehat{=}}$ use of integers, $a_3 \mathrel{\widehat{=}}$ use of natural numbers, $b_1 \mathrel{\widehat{=}}$ calculation of powers, $b_2 \mathrel{\widehat{=}}$ addition.

Both operations of B can be applied on the sets of numbers of A. Therefore, we can construct problems that contain one property of A and one property of B. The problem $(-5)^2$, for instance, corresponds to the combination of a_2 and b_1.

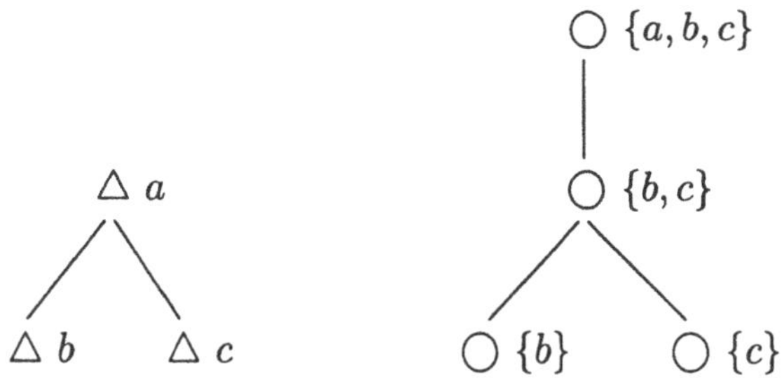

FIG. 4. Component structure and problem structure for Example 2.6.

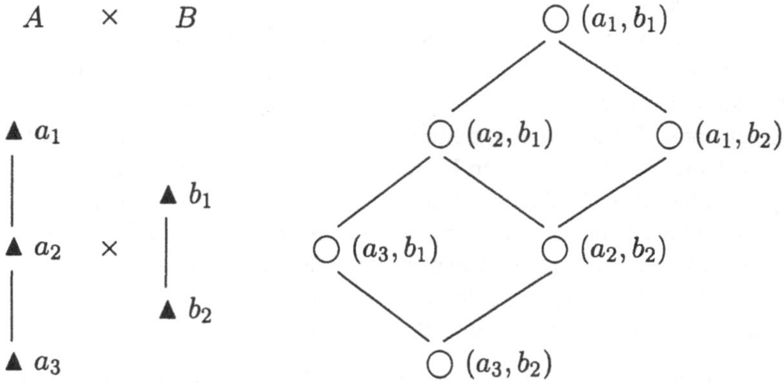

FIG. 5. Orders of attributes and problem structure based on componentwise ordering for Example 2.7.

From A and B we can construct a set $\mathcal{F}_p$ of six problems:

$$\mathcal{F}_p = \{(a_1, b_1), (a_1, b_2), (a_2, b_1), (a_2, b_2), (a_3, b_1), (a_3, b_2)\}.$$

We see that all problems of $\mathcal{F}_p$ consist of two components that are represented by their attributes $a_1, \ldots, a_i, \ldots, a_n$ and $b_1, \ldots, b_j, \ldots, b_m$. The problem structure for this example is shown in Fig. 5 (on the right-hand side). □

Let us now consider the principles by which these problems were constructed and ordered in detail. Components are sets of attributes. It is supposed that the attributes in a component cannot be combined with one another. In our example we have two components whose attributes are combined. This combination has been established by forming the *Cartesian product* of A and B. Looking at Example 2.7, we can easily check that $\mathcal{F}_p$ is a set that contains the product of A and B.

In order to establish a surmise relation as shown in Fig. 5, it is necessary to compare the generated problems in pairs with respect to the attributes of the components. Formally, the ordering rule we applied was:

> Let $C_1, \ldots, C_n$ be component sets on which partial orders $R_1, \ldots, R_n$ are defined. On the Cartesian product $C_1 \times \ldots \times C_n$ an order $\preceq$ is imposed by defining
>
> $$(x_1, \ldots, x_n) \preceq (y_1, \ldots, y_n) \iff (\forall i)\ x_i\ R_i\ y_i.$$

Expressed in words: We surmise that a problem q_1 is at least as difficult to solve as a problem q_2, if all attributes of q_1 are at least as difficult as the corresponding attributes in q_2 with respect to the relations R_i defined on the attribute sets. This principle is known as "coordinatewise order", for a description see Davey and

Priestley (1990, p. 18). According to Birkhoff (1973), $\preceq$ is a partial order (i. e. reflexive, transitive, and antisymmetric). Note that this method is also known from decision theory where the choice heuristic called *dominance rule* corresponds to coordinatewise orders.

Extensions of this ordering method applied to problems of elementary probability calculus are introduced in Held (1992, 1993). There, an approach to the component-based establishment of surmise systems can also be found.

Because the attributes of the components must be compared, it is necessary to define an order on each set of attributes. Example 2.7 showed the case of linearly ordered attributes. Example 2.8 demonstrates the ordering of problems that were constructed from quasi-ordered sets of attributes.

EXAMPLE 2.8 Let $A = \{a_1, a_2, a_3\}$ and $B = \{b_1, b_2, b_3\}$ be partial-ordered sets of attributes. Fig. 6 shows the Hasse diagrams for these sets and the corresponding problem structure. A procedure for the graphical construction of such products is given in Davey and Priestley (1990, p. 19). □

Problems that were constructed by product formation can also be ordered *lexicographically*.

EXAMPLE 2.9 As in Examples 2.7 and 2.8, we have two components $A = \{a_1, a_2, a_3\}$ and $B = \{b_1, b_2\}$. It is assumed that component A is "more important" than component B. Fig. 7 shows the lexicographic order of the product $A \times B$. How was this order established? First, we describe the general principle. The n-tuples that are to be ordered are compared pairwise beginning with the first elements (here: a_i). Because it is assumed that A is the most "important" component, it is also assumed that if these elements are not identical, the n-tuple that contains the subordinate element with respect to the order on A is subordinate to the other n-tuple. In Fig. 7 we see that this is the case for all tuples (a_1, b_i) and (a_2, b_j). If the first elements are identical, the second pair of elements will be compared and the n-tuple with the subordinate element is subordinate (see all tuples (a_i, b_1) and (a_i, b_2)). This procedure that is known from dictionaries continues on until two different elements are found or until there are no more elements left to compare. □

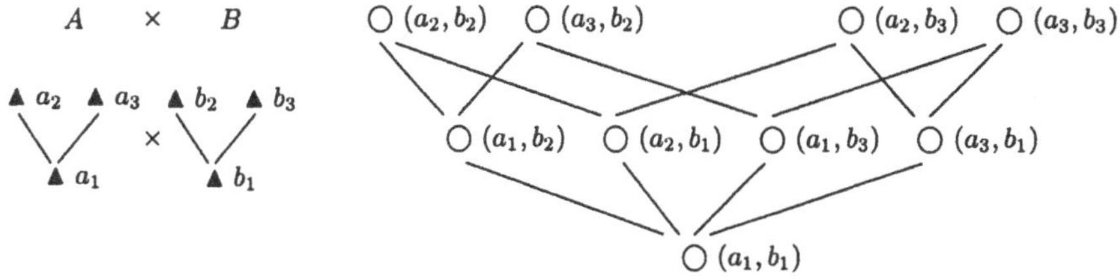

FIG. 6. Attributes and problem structure for Example 2.8.

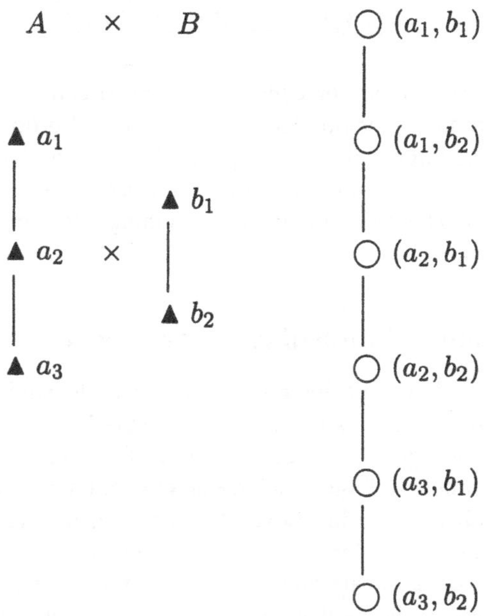

FIG. 7. Orders of attributes and problem structure based on lexicographic ordering.

For establishing a lexicographic order we need sets A_i of attributes and relations P_i that are defined on the sets A_i. $\langle A_i, P_i \rangle$ with $i = 1, \ldots, n$ have to be strict linear orders (i. e. irreflexive, transitive, and weakly connected). A lexicographic order is defined as follows:

DEFINITION 2.3 Let $\langle A_i, P_i \rangle, i = 1, \ldots, n$, be strict linear orders. $\langle A_1 \times A_2 \times \ldots \times A_n, L \rangle$ with $L \subseteq ((A_1 \times A_2 \times \ldots \times A_n) \times (A_1 \times A_2 \times \ldots \times A_n))$ is a *lexicographic order*, that is for

$$(a_1, a_2, \ldots, a_n), (b_1, b_2, \ldots, b_n) \in A_1 \times A_2 \times \ldots \times A_n$$

holds $(a_1, a_2, \ldots, a_n)\ L\ (b_1, b_2, \ldots, b_n)$ if and only if,

$$a_1 P_1 b_1 \ \vee\ (a_1 = b_1 \ \wedge\ a_2 P_2 b_2) \ \vee\ (a_1 = b_1 \ \wedge\ a_2 = b_2 \ \wedge\ a_3 P_3 b_3) \ \vee\ \ldots$$
$$\ldots \vee\ (a_1 = b_1 \ \wedge\ a_2 = b_2 \ \wedge\ \ldots \ \wedge\ a_{n-1} = b_{n-1} \ \wedge\ a_n P_n b_n)$$
$$\vee\ (a_1 = b_1 \ \wedge\ a_2 = b_2 \ \wedge\ \ldots \ \wedge\ a_n = b_n).$$

□

We introduced two important principles (i. e. set inclusion and componentwise ordering of product sets) for the construction of problems from components and for the establishment of a surmise relation on these problems. In the following section, we present two empirical investigations that make use of those principles.

EMPIRICAL EXAMPLES

The empirical examples report on experimental investigations which make use of the principles introduced for problem construction and problem ordering. The first investigation belongs to the area of psychology of problem solving. It deals with the solution of chess problems. In the second experiment, we focus on types of problems related to the field of inductive reasoning: the continuation of number series.

Construction and solution of chess problems

Chess playing is surely one of the most complex and demanding knowledge domains. This complexity makes the domain particularly interesting for cognitive scientists and psychologists. Not only the game of chess itself, but also the construction of chess problems requires a large amount of knowledge and experience. The immense number of possible moves that can be made, even from a very simple constellation, makes the decision, of whether one move is better than another very difficult. Grandmasters are often unable to "proof" what move is the best in a particular situation; therefore they often have to act intuitively.

An important book about the psychology of chess playing was written by De Groot (1965). He attempted to investigate the thought processes of highly trained chess players by means of introspective methods. De Groot also provided a proof scheme for objectively solvable positions, but the proof only works if someone is able to differentiate between "good" and "less good" moves. This differentiation has, for complex positions, to be intuitive.

We can already see that for the construction of chess problems, we should not attempt to focus on such demanding constellations that in addition to requiring highly evolved skills, are also very time consuming. In our example, we use the classical form of "three move problems" that are familiar to every chess player. In Fig. 8, we provide a typical example. The task is to perform the moves to reach a "winning position in three moves." Supposing white starts, the solution is: 1. Be2 h1Q; 2. Bh5+ Qh5:; 3. Ng7+. Experienced chess players can show that for this type of problem there is only one optimal solution. Furthermore, the time needed for handling such a position is expected to be much shorter than for a complex constellation in a real chess game.

As a next step, we have to find a way for the construction of such problems. Before we can apply one of our construction rules in the next section, components have to be introduced. A basic concept in chess playing are "*motives*", which are tactical standard situations. In terms of problem solving, motives can be seen as subgoals of a problem's solution. Fig. 9 shows examples for positions in which the motives "fork", "pin", "guidance", and "deflection" occur.[1] To illustrate we

[1] The example positions are identical to problems of the experimental investigation. Therefore

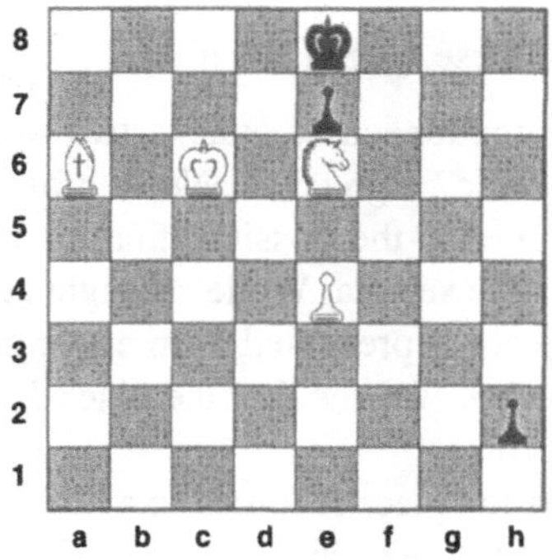

FIG. 8. A typical three move problem.

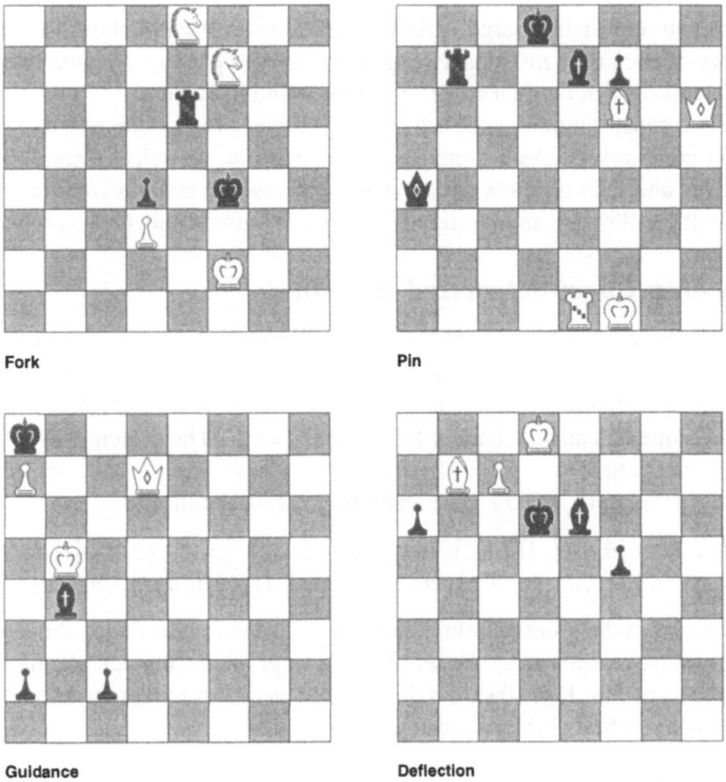

FIG. 9. Positions in which the motives "fork", "pin", "guidance", and "deflection" occur.

give a short description of these special situations:

- *Fork:* One piece simultaneously attacks two opposing pieces of higher value. Solution:[2] 1. Nc7 Rg6/c6; 2. Nd5+ arbitrary[3]; 3. Ne7+/Ne5+. If we take a look at one of the possible final positions (*Black:* Kf5, Rc6, ... *White:* Ne7, ...), we see that White's Knight attacks both Kf5 and Rc6.
- *Pin:* An opposing piece is prevented from moving. Solution: 1. Qf8+ Qe8; 2. Rd1+ Rd7; 3. Be7:+. We see that the black Bishop cannot move away from e7 because of Bf6+.
- *Guidance:* An opposing piece is forced to a disadvantageous square. Solution: 1. Kb6 Ba5+/c5+; 2. Ka6c6 arbitrary; 3. Qb7/c6 mate; Black's Bishop is forced to a5+/c5+, otherwise 2. Qb7 mate.
- *Deflection:* An opposing piece is forced to leave an important line or square. Solution: 1. Bc8 Bd5; 2. Bf5: Bb7; 3. Be4. Black's Bishop is forced to leave e6, otherwise 2. Be6: ...

Motives can appear in a large variety of combinations and belong to the basic repertory of even only moderately experienced chess players. A complete list of all problems used in our investigation can be found in Table 1.

For the construction of problems, these motives present one possible type of problem components. As a principle of construction, we select a small number of motives and then produce three move problems that contain *combinations* of them. In the following, an investigation that makes use of this idea, is reported.[4]

Problem Construction and Hypothesis

As we have already indicated, the construction of problems and the establishment of the surmise relation are based on the combination of motives. The motives—symbolized by a, b, c, and d—are elements of a single component set C. We assume that an antichain order is defined on C. The principle of set inclusion will be applied.

The component space $\mathcal{F}_C$ (see Definition 2.2) is as follows:

$$\begin{aligned}\mathcal{F}_C = \; & \{\emptyset, \{a\}, \{b\}, \{c\}, \{d\}, \{a,b\}, \{a,c\}, \{a,d\}, \{b,c\}, \{b,d\}, \\ & \{c,d\}, \{a,b,c\}, \{a,b,d\}, \{a,c,d\}, \{b,c,d\}, \{a,b,c,d\}\}.\end{aligned}$$

By means of the ordering principle of set inclusion as introduced above we can infer a surmise relation R on the set Q of the 15 problems that are identified with the elements of $\mathcal{F}_C$. Fig. 10 shows this relation as a Hasse diagram. Expressed in words, the hypothesis for the investigation is:

these examples may appear to be rather complex.

[2] The "solution" provides the sequence of three moves which a chess expert has considered as optimal for reaching a winning position.

[3] "Arbitrary" means that this move (in this case Black's move) is not relevant to the solution.

[4] The investigation was conducted by B. Hierholz at the University of Heidelberg under direction of the first author.

TABLE 1
Complete List of Chess Problems.

Number	Type	Position	Solution	Motives
1	abcd	*White*: Ka7 Qh3 Re5 Nd6 *Black*: Kh8 Qg6 Rg8 Bf7 Ph7	1. Rg5 Qf6 2. Qc3 Qc3: 3. Nf7 mate	deflection, guidance, pin, fork
2	bcd	*White*: Kh2 Bf3 Nh5 Pg3,g7 *Black*: Kh7 Qe6 Ph3	1. g8Q+ Kg8: 2. Bd5 Qd5: 3. Nf6+	guidance, pin, fork
3	abc	*White*: Kg1 Qe2 Re1 Bg6,h2 Pf2 *Black*: Kf8 Qb7 Rg8 Be7,h3 Pg7,f6	1. Qe7:+ Qe7: 2. Bd6 Qd6: 3. Re8 mate	deflection, guidance, pin
4	acd	*White*: Kg1 Qc2 Rf2 Bb1 Nf8 Pb2,c6,g2 *Black*: Kd8 Qg7 Rd6 Be4 Nd3 Pb7,e7	1. cb: Bb7: 2. Qd3: Rd3: 3. Ne6+	deflection, pin, fork
5	abd	*White*: Ka2 Qf4 Be3 Pb2,b3,h3,c7 *Black*: Ka5 Qe7 Nb6 Pa6,b5,c5,b4,h4	1. Qb4:+ cb: 2. Bb6:+ Kb6: 3. c8N+	deflection, guidance, fork
6	bc	*White*: Kf1 Qa6 Re1 Nh3 Pg2,f2,d4 *Black*: Ke8 Qd6 Rh8 Nc6 Pe6,f7,g7	1. d5 Qd5: 2. Qa8+ arbitrary 3. Qc6:+/Qd5:/Qh8:	guidance, pin
7	ad	*White*: Kd6 Nf5 Pe7 *Black*: Kf7 Ng4 Ph7	1. Nh6+ Nh6: 2. Ke2 arbitrary 3. e8Q	deflection, fork
8	bd	*White*: Kc6 Ba6 Ne6 Pe4 *Black*: Ke8 Pe7,h2	1. Be2 h1Q 2. Bh5+ Qh5: 3. Ng7+	guidance, fork
9	ac	*White*: Kh2 Bb6 Pf3,g2 *Black*: Kh4 Rc2 Ph7,h5,g5	1. Bc7 Rg2:+ 2. Kg2: arbitrary 3. Bd8/f2 mate	deflection, pin
10	cd	*White*: Kf3 Rc6 Ne5 Pg5 *Black*: Kg8 Rd4 Be7 Pf4	1. Rc8+ Kg7 2. Rc7 Kf8 3. Ng6+	pin, fork
11	ab	*White*: Kh2 Qd1 Re2 Pd7,f2,h4 *Black*: Kg8 Qb5 Rd8 Pa4,g7,h7	1. Re8+ Re8: 2. Qd5+ Qd5: 3. deQ mate	deflection, guidance
12	d	*White*: Kf2 Ne8,f7 Pd3 *Black*: Kf4 Re6 Pd4	1. Nc7 Rg6/c6 2. Nd5+ arbitrary 3. Ne7+/Ne5+	fork
13	c	*White*: Kf1 Qh6 Re1 Bf6 *Black*: Kd8 Qa4 Rb7 Be7 Pf7	1. Qf8+ Qe8 2. Rd1+ Rd7 3. Be7:+	pin
14	b	*White*: Kb5 Qd7 Pa7 *Black*: Ka8 Bb4 Pa2,c2	1. Kb6 Ba5+/c5+ 2. Ka6/c6 arbitrary 3. Qb7/c6 mate	guidance
15	a	*White*: Kd8 Bb7 Pc7 *Black*: Kd6 Be6 Pf5,a6	1. Bc8 Bd5 2. Bf5: Bb7 3. Be4	deflection

Problem 8 by Maiselis and Judowitsch (1966); problem 10 by Geisdorf (1984); problem 12 by Chéron (1960); problem 14 by Speckmann (1958); the other problems by B. Hierholz.

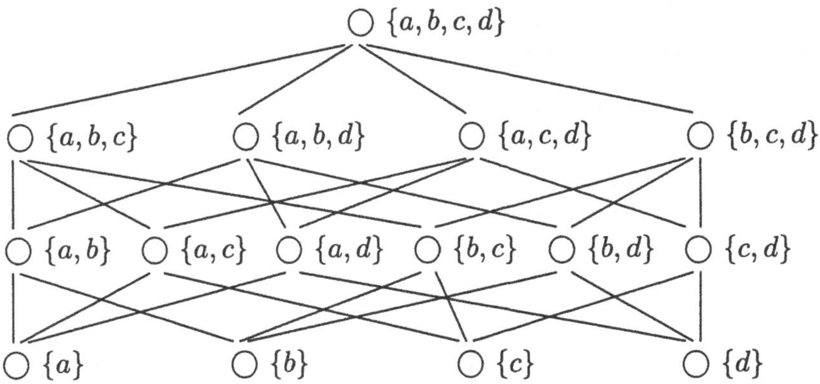

FIG. 10. Hasse diagram for the problems identified with the elements of the component space $\mathcal{F}_C$.

> *If a problem q identified with a component set $T \in \mathcal{F}_C$ is solved by a subject, then all problems q' which are identified with a component set $T' \in \mathcal{F}_C$ with $T' \subseteq T$ will also be solved by this subject.*

The hypothetical structure corresponds to a set of 167 knowledge states. This means that only about 0.5 % of $2^{15} = 32768$ potential solution patterns represent valid states.

Method

For the investigation, four motives were selected and combined as shown in Fig. 10. These motives are "fork", "pin", "deflection" and "guidance." The combinations of these four motives form a set of 15 problems. The complete list of the problems is given in Table 1. The positions of Fig. 9 are examples for the problems with only one motive. Fig. 8 presents a problem with the two motives, "guidance" and "fork." These problems were presented to 13 subjects who were all members of the chess club in Ladenburg, Germany.

First, the subjects were asked to read the instructions for the experimental procedure, then they were permitted to begin working on the problems. Each problem was printed on a single card as a diagram (see Fig. 8 and 9). The subjects had to write down the solution in the usual form used above.

The time needed for the solution was controlled by the subjects themselves with the aid of a chess clock. There was no time limit. The subjects were asked only to answer "as accurately and as quickly as possible." The problems were presented in the order of hypothesized difficulty, so problem $\{a, b, c, d\}$ with four motives was the first to be presented and the one-motive problems $\{a\}$, $\{b\}$, $\{c\}$, and $\{d\}$ were the last to be presented.

Results

The criteria for the goodness of fit that will be used in this paper are (1) the *minimal distances* between each solution pattern and the closest states in the quasi-ordinal knowledge space and (2) the *mean distance* between the set of solution patterns and the states in the quasi-ordinal knowledge space. Let $\mathcal{K}$ be a quasi-ordinal knowledge space, and let $\mathcal{S}$ be the set of all solution patterns in the data; let the elements of $\mathcal{K}$ as well as those of $\mathcal{S}$ be represented in the form of subsets of problems. Further, let $X \in \mathcal{S}$ and $K \in \mathcal{K}$. Then the *distance* between X and K, abbreviated by $dist(X, K)$, is defined as the number of elements occurring in the symmetric set difference of X and K, that is,

$$dist(X, K) = |(K \setminus X) \cup (X \setminus K)|.$$

The *minimal distance* of X to $\mathcal{K}$, abbreviated by $mdist(X, \mathcal{K})$, is defined as the distance of X to the "nearest" knowledge state in $\mathcal{K}$, that is,

$$mdist(X, \mathcal{K}) = \min \{dist(X, K) | K \in \mathcal{K}\}.$$

Now, the *mean distance* $d(\mathcal{S}, \mathcal{K})$ of $\mathcal{S}$ to $\mathcal{K}$ is given by

$$d(\mathcal{S}, \mathcal{K}) = \sum \{mdist(X, \mathcal{K}) | X \in \mathcal{S}\} / N,$$

where N is the number of response patterns in $\mathcal{S}$. Table 2 shows the frequencies of minimal distances and and the mean distance d for the solution patterns of the 13 subjects and the 167 knowledge states of the hypothetical knowledge space.

The hypothesis holds only for three subjects who solved all problems and for one subject who failed only in solving problem $\{a, b, c, d\}$. Two subjects out of 13 each show inconsistencies for only one problem.

Discussion

The results clearly contradict our deterministic hypothesis because the response patterns of only four subjects agree with it. The reasons for the unsatisfactory results may be found both in the theoretical approach and the experimental design. First of all, the difficulty of the chess problems is probably not solely

TABLE 2
Chess Problems: Minimal Distances and Mean Distance Between the Solution Patterns and the Hypothetical Knowledge Space

	Distances							*d*
	0	1	2	3	4	5	6	
Frequencies	4	2	1	3	2	0	1	2.08

influenced by the type and number of included motives. An investigation by Albert, Schrepp, and Held (1994) show that taking the sequence of motives within problems into consideration can contribute to a more adequate problem structure.

Another problem common to investigations dealing with chess playing is that the work on chess problems requires great concentration over a large period of time. Thus, we suspect that the order of problem presentation (beginning with $\{a, b, c, d\}$) might not have been the best choice. The experimental setting as a group experiment and the lack of any limit on solution times may have caused a decrease in motivation with some of the subjects who required longer solution times.

In the investigation of Albert, Schrepp, and Held (1994) mentioned above, these problems were taken into account. A computerized experimental laboratory setting was used. Further, the uniqueness of motive assignment was optimized. Due to these improvements the results of this investigation are much more conclusive than the ones reported here.

Continuing a series of numbers

Our second empirical example deals with a type of task that is commonly found in diagnostic instruments in psychology. It is typical for inductive reasoning. A series of numbers constructed according to an algebraical rule is to be continued by one or more numbers. Subjects are required to infer the rule from the number series presented and to calculate the missing number with the help of this rule. The following example demonstrates a very simple task:

30 32 36 44 60 ... ?

One possible rule is: $x_n = x_{n-1} + 2^n$. Of course, we can find other formulas that correspond to the example, e. g. $x_n = 3x_{n-1} - 2x_{n-2}$, where x_n is the number, we are trying to find, x_{n-1} is the preceding number (here: 60), and so on. Our example shows that both formulas use preceding elements of the given series for the calculation of x_n. We call the number of immediate predecessors that are used for the solution of the problem the *level of recursion*. The first formula has recursion level "1", the second is of level "2." Krause (1985) used this type of recursively connected number series in an investigation of mental processes and rule detection. He attempted to classify the various methods subjects used to solve this type of problem.

Some types of number series problems possess properties that make them suitable for our component-based method of problem construction. The level of recursion is one of them. Generally, we assume that the following cognitive demands are covered by number series problems: (1) the subject has to recognize properties and regularities of the presented sequence (e. g. the level of recursion),

TABLE 3
Number Series: Problem Components

Components		*Attributes*	
$\mathbf{M_1}$	$\mathbf{a_1}$ level of rec.: 3	$\mathbf{a_2}$ level of rec.: 2	$\mathbf{a_3}$ level of rec.: 1
$\mathbf{M_2}$	$\mathbf{b_1}$ multiplicative factor $f > 1 \wedge f \in N$	$\mathbf{b_2}$ multiplicative factor $f = 1$	
$\mathbf{M_3}$	$\mathbf{c_1}$ additive factor $g > 1 \wedge g \in N$	$\mathbf{c_2}$ additive factor $g = 0$	

(2) a hypothesis concerning the underlying rule has to be established, applied, and tested.

Problem construction and hypothesis

Number series problems are extremely variable, so the question is, what types of components can be combined in which ways. In this investigation, three distinct components M_1, M_2, M_3 were used. Their attributes are shown in Table 3. Concerning attributes b_2 and c_2, we must note that the definition of the factors $f = 1$ and $g = 0$ is included for "technical" reasons: although a recognition of a multiplicative or additive factor is not necessary for a solution of the problems which are characterized by b_2 or c_2, giving such "zero values" is appropriate for a complete problem definition by elements of a Cartesian product. We assume that a linear order is defined on the attributes of each component. The Hasse diagrams of Fig. 11 (left) illustrate this fact. This assumption means that, for example, recursion level 3 makes a problem more difficult than recursion level 2, or the existence of a multiplicative factor that is greater than 1 provides more complication than factor 1 that psychologically corresponds to no demand for detecting this factor.

Now we must define a problem construction rule for these components. In the previous section, we demonstrated how product formation can be applied to sets of components. We apply this rule to M_1, M_2, M_3. The product $M_1 \times M_2 \times M_3$ provides twelve combinations of attributes of the type (a_n, b_n, c_n). We call the set of these combinations *problem set* Q_t.

The next step is the application of the *componentwise ordering rule*. This leads to the structure of the twelve problems, where problem (a_1, b_1, c_1) is assumed to be the most difficult and (a_3, b_2, c_2) the simplest. Table 4 shows the complete problem set constructed for the investigations. On the right side of Fig. 11, we can see the problem structure.

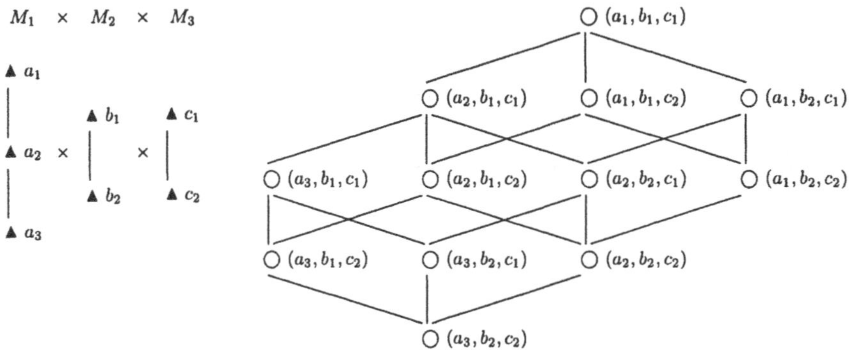

FIG. 11. Number series: Orders of attributes and problems.

TABLE 4
Number Series: Calculation Rules, Problems, and Solutions

Attributes		c_1	c_2
a_1	b_1	$x_n = 2x_{n-3} + x_{n-2} + x_{n-1} + 4$ $1, 5, 9, 20, 43, 85 \Rightarrow 172$	$x_n = 2x_{n-3} + x_{n-2} + x_{n-1}$ $6, 6, 7, 25, 44, 83 \Rightarrow 177$
	b_2	$x_n = x_{n-3} + x_{n-2} + x_{n-1} + 1$ $16, 16, 17, 50, 84, 152 \Rightarrow 287$	$x_n = x_{n-3} + x_{n-2} + x_{n-1}$ $26, 34, 41, 101, 176, 318 \Rightarrow 595$
a_2	b_1	$x_n = x_{n-2} + 2x_{n-1} + 2$ $1, 4, 11, 28, 69, 168 \Rightarrow 407$	$x_n = 2x_{n-2} + x_{n-1}$ $5, 11, 21, 43, 85, 171 \Rightarrow 341$
	b_2	$x_n = x_{n-2} + x_{n-1} + 5$ $12, 17, 34, 56, 95, 156 \Rightarrow 256$	$x_n = x_{n-2} + x_{n-1}$ $25, 34, 59, 93, 152, 245 \Rightarrow 397$
a_3	b_1	$x_n = 2x_{n-1} + 1$ $7, 15, 31, 63, 127, 255 \Rightarrow 511$	$x_n = 2x_{n-1}$ $4, 8, 16, 32, 64, 128 \Rightarrow 256$
	b_2	$x_n = x_{n-1} + 13$ $33, 46, 59, 72, 85, 98 \Rightarrow 111$	$x_n = x_{n-1}$ $113, 113, 113, 113, 113 \Rightarrow 113$ (not presented in Investigation I)

Method

The problems used in the investigations have the following properties: the multiplicative constant is either 2 or 0, the additive constant is always a single- or two-digit element of $\mathbf{N}_0$, and the maximal recursion level is 3. To avoid successful guessing of the solution, all solutions are numbers greater than 100. Table 4 shows all calculation rules and the corresponding problems. The problems were used in two investigations. Investigation I took place at the Department of Psychology of the University of Heidelberg, Germany, with 18 subjects. Investigation II was conducted with 30 subjects at the Department of Psychology of the University of Graz, Austria.[5]

In Investigation II, the complete set of 12 problems was presented, where as in Investigation I the trivial problem (a_3, b_2, c_2) was not used assuming that it would possibly have confused the subjects.

The hypothetical structure $\mathcal{K}_1$ for 11 problems (as used in Investigation I) corresponds to 49 knowledge states. This is 2.4 % of $2^{11} = 2048$ possible solution patterns. In the case of 12 problems (Investigation II), we have a hypothetical structure $\mathcal{K}_2$ with 50 knowledge states (i. e. 1.2 % of $2^{12} = 4096$ possible patterns).

In both investigations, the subjects were asked to read instructions that introduced the problem type. The subjects were also told that the only mathematical operations to be used were addition and multiplication with non negative numbers. Then they were asked to solve three simple example problems.

The problems were then presented (printed on cards) in a randomized order. The subjects had to write down the solution on a sheet of paper. If the solution was not given within 7 minutes, the next problem was presented. Subjects who either gave the wrong solution or wanted to "give up" before the 7 minutes had expired were asked to go on thinking about the problem. After the last problem, the subject was asked for the rules he or she used for solving the problems.

Results[6]

None of the subjects in Investigation II failed solving the trivial problem (a_3, b_2, c_2). Therefore, we ignore this problem in our further analysis and consider only knowledge space $\mathcal{K}_1$ for the comparisons between data and hypothetical structure.

In Table 5, the results of the two investigations are presented separately. In addition the results for the whole group of 48 subjects are shown. We see that in Investigation I the solution pattern of 16 out of 18 subjects is identical with a knowledge state in $\mathcal{K}_1$. Only patterns of two subjects have a distance of 1 to a state in $\mathcal{K}_1$. In Investigation II, solution patterns with distances of 2 and 3 also occurred. All in all, 14 different solution patterns with a distance of 0 have been observed.

Discussion

The results of this investigation show that the hypothetical conclusions we drew about the componentwise ordering rule were rather accurat.

[5] Investigation I was conducted by P. Hellriegel, J. Ptucha and M. Wölk. Investigation II was conducted by Birgit Edlinger and Dagmar Spreitzhofer. Both investigations were guided by the first author.

[6] We are grateful to Jochen Musch (University of Bonn), who conducted a first analysis of the data of Investigation II.

TABLE 5
Number Series: Minimal Distances and Mean Distance Between the Solution Patterns and the Hypothetical Knowledge Space for Investigation I and Investigation II and the Whole Group of Subjects

	Distances				*d*
	0	1	2	3	
Investigation I (18 subjects)	16	2	—	—	.11
Investigation II (30 subjects)	18	7	3	2	.63
Both investigations (48 subjects)	34	9	3	2	.44

TABLE 6
Number Series: Minimal Distances and Mean Distance Between the Solution Patterns and the Knowledge Space that is Based on a *Lexicographic Order* for Investigation I and Investigation II and the Whole Group of Subjects.

	Distances				*d*
	0	1	2	3	
Investigation I (18 subjects)	4	11	2	1	1.00
Investigation II (30 subjects)	6	12	7	5	1.37
Both investigations (48 subjects)	10	23	9	6	1.23

An alternative and "more economical" theory for the data could be stated by a lexicographic order on the problem set. For this purpose, we assume that component M_1 (recursion level) is the 'most important' component, M_2 (multiplicative factor) the second most important, and M_3 the least important component. With respect to this order, only 4 solution patterns of Investigation I and 6 patterns of Investigation II agree with a state. Table 6 provides an overview. In this case, only 12 knowledge states (consisting of 11 problems) are assumed to exist—these are about .6 % of the potential response patterns. Although the lexicographic order is much more "restrictive" than the componentwise order, these results may also be a product of the assumption concerning the importance of the components.

We assumed that a subject who is able to solve a problem, will use one particular calculation rule. This is not always realistic because, for every number series problem, alternative solutions can be found. These alternatives are frequently also plausible. The next section deals with alternative solutions.

Ambiguity of Number Series Problems

In correspondence with problem construction, the calculation rule for the series $5, 11, 21, 43, 85, 171, \ldots ?$ is $x_n = 2x_{n-2} + x_{n-1}$. This is problem (a_2, b_1, c_2). Obviously, the rule $x_n = 2x_{n-1} + (-1)^n$ will also provide a correct solution.

Although the subjects were told that only positive constants are to be added in the problems, we cannot exclude the possibility that a subject will use such an alternative rule. In the reported Investigation I, eleven subjects provided a correct answer, whereby eight subjects used the alternative rule as shown above.

We can see that the construction and ordering of number series problems must be based on an exact analysis of the uniqueness of the problems, especially if those are constructed from components that include principles of solution. Korossy (1998) examined the phenomenon of ambiguous number series problems with special reference to the case of linear recursive series. He developed a method, that allows the uniqueness of the solution to be determined. This method is based on the theory of linear equation systems. One of the main results of his study is that only heavy restrictions on the domains of the recursive formulas lead to less ambiguous ranges for the solutions. In the case of different rules for a problem, a generalized model using surmise systems and knowledge spaces and the respective principles for constructing these structures (see Held, 1993; Held, Chapter 4, this volume) may be appropriate.

As an overall conclusion, we can say that it is impossible to construct a number series problem that can be solved by only one rule. However it is possible to minimize the number of alternatives to a degree that allows one to work with this type of problem. Furthermore, if the manner in which an ambiguous problem has been solved is known, it may be possible to infer which of the assumed cognitive demands has been mastered by the subject.

GENERAL DISCUSSION

In this chapter, methods for the generation of ordered problem sets are introduced. Our theoretical results are motivated by the theory of knowledge spaces (Doignon & Falmagne 1985). A basic concept of this theory is the *surmise relation*, a transitive and reflexive binary relation defined on a set of problems. By this relation, a set of *knowledge states* (i. e. subsets of the problem set) is determined. Although the step from surmise relations to the more general concept of *surmise systems* is the main achievement of the theory of knowledge spaces, we restrict our considerations concerning this theory to surmise relations.

The question we are focusing on is how surmise relations can be derived from a systematically constructed set of problems. Both problem construction and problem ordering are based on *domain specific theories*, which are prerequisites for the definition of *problem components* and the establishment of *problem structures* that are derived from these components. Problem components may, for example, be operations necessary for a problem solution or subgoals during the solution process.

The methods introduced for the establishment of ordered problem sets are in principle known from elementary ordering theory and are well known in psychology: *set inclusion* and *componentwise ordering of product sets*.

We present two applications of the methods introduced for problem construction and problem ordering. The first investigation deals with the domain of *problem solving*. Chess problems are constructed on the basis of *motives*. These tactical elements of the game of chess are viewed as subgoals for the solution process that have to be detected as well as realized. A surmise relation on the problem set is established by inclusion of motive sets. The second investigation belongs to the domain of *inductive reasoning* with the solution of number series problems. Problem construction is done by product formation. The surmise relation is a result of the componentwise ordering of products. In this case, the components are parts of the *rules* that have to be found for problem solution.

Further principles for the establishment of knowledge structures that are based on problem components or skills have been developed. Lukas and Micka (1993) considered the assignment of skills to elementary chess-endgame problems. In Lukas (1997), the solution of problems on basic electricity circuits is modeled by *information systems*. This approach also focuses on *incompatibility relations* between skills. These results are also important for the definition of component-based problems as introduced in this article.

The investigations of Korossy (1993, 1997) are based on modeling *competencies* and *performances* within assessment processes. The domain under investigation is the field of *geometric constructions and calculations*. In Held (1992, 1993), knowledge spaces are derived from component-based problems on *elementary combinatorics and probability calculus*. Some of the theoretical approaches introduced there are extensions of the methods of this chapter (e. g., principles for constructing surmise systems and surmise relations). Furthermore, the assignment of "problem demands" to problem components is discussed.

Albert, Schrepp and Held (1994) provided the principle of *sequence inclusion* for ordering motive-based chess problems. This method is an extension of set inclusion that has been used for the chess experiment reported here.

ACKNOWLEDGEMENTS

The research reported in this paper is based on Albert (1989, 1991) and Albert and Held (1994); it was supported by Grant Lu 385/1 of the Deutsche Forschungsgemeinschaft to J. Lukas and D. Albert at the University of Heidelberg. We are grateful to J. Heller (University of Regensburg), J. Lukas (University of Halle), and H. Rodenhausen (University of Köln) for their invaluable comments on an earlier draft of this paper.

REFERENCES

Albert, D. (1989) *Knowledge assessment: Choice heuristics as strategies for constructing questions and problems.* Paper read at the 20th European Mathematical Psychology Group Meeting, Nijmegen, September 1989.

Albert, D. (1991) *Principles of problem construction for the assessment of knowledge.* Paper read at the 2nd European Congress of Psychology, Budapest, July 1989.

Albert, D., & Held, T. (1994). Establishing knowledge spaces by systematical problem construction. In D. Albert (Ed.), *Knowledge structures* (pp. 81–115). Berlin, Heidelberg: Springer.

Albert, D., Schrepp, M., & Held, T. (1994). Construction of knowledge spaces for problem solving in chess. In G. Fischer & D. Laming (Eds.), *Contributions to mathematical psychology, psychometrics, and methodology* (pp. 123–135). New York: Springer.

Birkhoff, G. (1937). Rings of sets. *Duke Mathematical Journal, 3,* 443–454.

Birkhoff, G. (1973). *Lattice theory* (3rd ed.). Providence: American Mathematical Society.

Chéron, A. (1960). *Lehr- und Handbuch der Endspiele Bd. 1.* [Text- and handbook of endgames Vol. 1.] Berlin: S. Engelhardt-Verlag.

Davey, B. A., & Priestley, H. A. (1990). *Introduction to lattices and orders.* Cambridge: Cambridge University Press.

De Groot, A. D. (1965). *Thought and choice in chess.* The Hague: Mouton & Co.

Doignon, J.-P., & Falmagne, J.-C. (1985). Spaces for the assessment of knowledge. *International Journal of Man-Machine Studies, 23,* 175–196.

Doignon, J.-P., & Falmagne, J.-C. (1998). *Knowledge spaces.* Berlin: Springer.

Falmagne, J.-C., Koppen, M., Johannesen, L., Villano, M., & Doignon, J.-P. (1990). Introduction to knowledge spaces: How to build, test and search them. *Psychological Review, 97*(2), 201–224.

Geisdorf, H. (1984). *Der Schachfreund, Bd. 1: Auf den Flügeln der Kunst.* [The chess buff.] Mannheim: Selbstverlag H. Geisdorf.

Guttman, L. A. (1947). A basis for scaling qualitative data. *American Sociological Review, 9,* 139–150.

Guttman, L. A. (1950). The basis for scalogram analysis. In S. A. Stouffer, L. A. Guttman, E. A. Suchman, P. F. Lazarsfeld, S. A. Star & J. A. Clausen (Eds.),*Studies in social psychology in world war II, Volume 4: Measurement and prediction* (pp. 60–90). London: Princeton University Press.

Held, T. (1992). *Systematische Konstruktion und Ordnung von Aufgabenmengen zur elementaren Wahrscheinlichkeitsrechnung.* [Systematical construction and ordering of problem sets of elementary stochastics.] Paper read at the 34th Tagung experimentell arbeitender Psychologen, Osnabrück, FRG, 12.-16. April 1992.

Held, T. (1993). *Establishment and empirical validation of problem structures based on domain specific skills and textual properties—A contribution to the "Theory of knowledge spaces".* Unpublished doctoral dissertation, University of Heidelberg, Germany.

Korossy, K. (1993). *Modellierung von Wissen als Kompetenz und Performanz. Eine Erweiterung der Wissensstruktur-Theorie von Doignon & Falmagne.* [Modeling knowledge as competence and performance. An extension of Doignon and Falmagne's theory of knowledge spaces.] Unpublished doctoral dissertation, University of Heidelberg, Germany.

Korossy, K. (1997). Extending the theory of knowledge spaces: A competence–performance approach. *Zeitschrift für Psychologie, 205*, 53–82.

Korossy, K. (1998). Solvability and uniqueness of linear-recursive number sequence tasks. *Methods of Psychological Research Online, 3*(1), 43–68.

Krause, B. (1985). Zum Erkennen rekursiver Regularitäten. [Concerning detection of recursive regularities.] *Zeitschrift für Psychologie, 193*, 71–86.

Lukas, J. (1997). Modellierung von Fehlkonzepten in einer algebraischen Wissensstruktur. [Modeling misconceptions in an algebraic knowledge structure.] *Kognitionswissenschaft, 4*, 196–204.

Lukas, J., & Micka, R. (1993) *Zur Diagnose von Wissen über einfache Schachendspiele: Formale Theorie und empirische Ergebnisse.* [Diagnosis of knowledge about simple chess endgames: formal theory and empirical results.] Paper read at the 35th Tagung experimentell arbeitender Psychologen, Trier, 4.-8. April 1993.

Maiselis, I. L., & Judowitsch, M. M. 1966). *Lehrbuch des Schachspiels.* [Textbook of chess.] Berlin: Sportverlag.

Speckmann, W. (1958). *Strategie im Schachproblem.* [Strategy in chess problem.] Berlin: de Gruyter.

3

Component-based Construction of Surmise Relations for Chess Problems

Martin Schrepp
Ruprecht-Karls-Universität Heidelberg

Theo Held
Martin-Luther-Universität Halle-Wittenberg

Dietrich Albert
Karl-Franzens-Universität Graz

An approach to construct surmise relations or quasi–ordinal knowledge spaces through ordering principles is described. The ordering principles apply to the components of problems, which are considered as the basic units of knowledge necessary to solve the problems properly. We describe the three ordering principles "set inclusion", "multiset inclusion" and "sequence inclusion" and an application of these principles to the construction of surmise relations on sets of chess problems. The basic units for the construction of surmise relations in chess are the tactical elements of the game—the "motives". In terms of problem solving, these motives can be regarded as subgoals in the process of problem solving. The empirical validity of the described ordering principles is tested in two experimental investigations. The results show that the two principles "multiset inclusion" and "sequence inclusion" predict the difficulty of chess problems rather well, whereas the principle "set inclusion" is clearly insufficient in this field. The experimental investigations also demonstrate the suitability of the theory of knowledge spaces for testing psychological theories.

INTRODUCTION

A crucial problem in the theory of knowledge spaces is how to establish a knowledge space on an item set. This may be done in several ways. First, there is the possibility of querying experts. Questioning procedures have been developed by Dowling (1991, 1993), Koppen and Doignon (1990), and Koppen (1993). Another approach is to analyze response patterns as proposed by Airasian and Bart (1973), Bart and Krus (1973), and Van Leeuwe (1974). A third approach—the one used in our investigations—is to infer the knowledge space from ordering principles or skill assignments, which apply to the components of the problems. This approach is also described in Albert and Held (1994), Albert, Schrepp, and Held (1994), Doignon (1994), Held (1993), and Korossy (1993).

The ordering principles *set inclusion*, *multiset inclusion*, and *sequence inclusion*, which are described in the following section in detail, enable us to construct surmise relations on problem sets, that is they always lead to quasi-ordinal knowledge spaces. We apply these principles in order to construct surmise relations in the domain of chess.

Chess involves one of the most complex and demanding knowledge domains. Both the game of chess itself and also the construction of problems that will serve to assess knowledge concerning the tactical elements of chess require much knowledge and experience and—as we show here—some special principles.

What types of problems might prove suitable for assessing a player's knowledge of chess? For which types of problems within the large domain of chess playing is experimental exploration feasible? Because chess is so very complex we restrict considerations to problems with unique solutions. In addition, we need to be able to specify the particular knowledge of chess, that is needed to solve each problem.

Tactical chess problems fulfill these requirements. Two typical problems of this kind are shown in Fig. 1. In both positions White moves first, and the problem is to find the best moves for White. The solution of problem (a) is: 1. Ktd5+ cxd5; 2. Bg3 Qxg3 stalemate. The solution of problem (b) is: 1. Bf4 Qxf4; 2. Ktd5+ K˜ ; 3. Ktxf4 draw.

Solving such problems requires knowledge of tactical elements, that are called "motives" in chess terminology. We try to use such motives for a classification of the problems' difficulty with respect to the other elements of the problem set. The motives considered in our investigations are "fork", "guidance", "elimination", "clearing", "promotion" and "stalemate". Definitions of these motives are given in Table 1.

We now provide an example of how a chess problem can be coded as a sequence of motives. The goal of White in problem (a) is to achieve a draw by forcing stalemate. To achieve this, White must eliminate the Knight on c3. The move 1. Ktd5+ forces the elimination (cxd5) of the Knight. The move 2. Bg3 forces the black Queen to the disadvantageous square g3 by Qxg3, and reaches

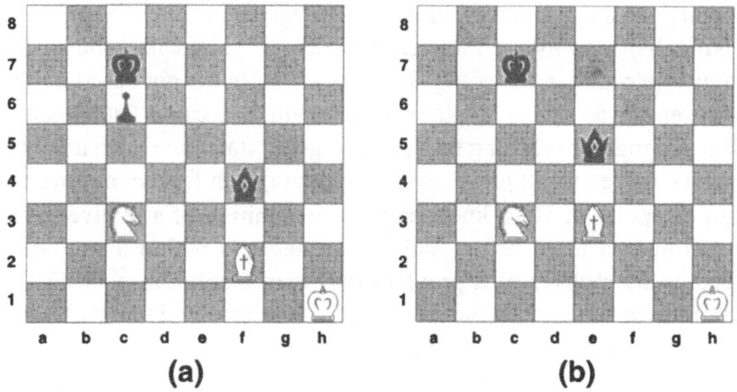

FIG. 1. Two examples for tactical chess problems.

TABLE 1
Description of Motives

Motive	*Name*	*Description*
F	Fork	One piece simultaneously attacks two opposing pieces of higher value.
G	Guidance	An opposing piece is forced to a disadvantageous square.
E	Elimination	The elimination of an own piece is forced, e.g., to achieve a stalemate.
C	Clearing	An important line or square is left, e.g., to achieve a stalemate.
T	Promotion	A pawn has to be promoted to the most suitable piece, by moving to the 8th line.
S	Stalemate	A stalemate has to be anticipated and has either to be provoked or avoided.

a draw by stalemate. Hence, the three motives "elimination", "guidance" and "stalemate" are contained in problem (a) in this sequence. The goal in problem (b) is (by winning the black Queen) to reach a draw. The move 1. Bf4 forces the black Queen to the disadvantageous square f4 (Qxf4) and the 'fork' 2. Ktd5+ wins the black Queen. Therefore, the motives "guidance" and "fork" are contained in problem (b) in this sequence.

As can be seen from these examples, the same move Ktd5+ in nearly identical positions is related to different motives. In problem (a) it is based on the motive "elimination", in problem (b) on the motive "fork". Hence, motives are mainly considered as "standard ideas" or "subgoals in the problem solving process". Note that we only use motives that occur as subgoals in the solution plan. For example in problem (a) of Fig. 1, the move 2. Bg3 is not based on the motive "pin" (preventing an opposing piece from moving), because "guidance", and not "pin" is a subgoal for reaching the terminal goal "stalemate". That is because, in this case, the Queen is not prevented from moving, but forced to move to g3.

From our point of view, knowledge or recognition of a motive means that a motive has been learned before and can be detected within a solution process. Recognizing the motives of a problem results in a solution plan. The solution process for these types of problems is therefore determined mainly by the combination and sequence of motives within a problem.

Motives are surely elements of the knowledge a person must possess if she or he is to be capable of solving the problems used in our investigations. But, there are probably other factors involved in a chess problem that also have an influence on the solution process. These factors are not the subject of our investigation; considerations regarding these factors can be found in the general discussion below.

We next introduce the theoretical principles underlying our investigations. We show how surmise relations may be established on sets of component-based chess problems. These surmise relations are inferred from psychological assumptions which are described below. These relations, the corresponding sets of knowledge states, constitute the hypotheses of our empirical investigations. The test of a hypothesis will therefore consist mainly of comparing the set of theoretically inferred knowledge states with the response patterns observed in an experiment.

THEORY

In this section, we describe three principles[1] for the component based establishment of surmise relations on sets of chess problems. These principles are based on the idea that the difficulty of a chess problem depends mainly on the motives that a subject must recognize in order to find the correct solution of the problem.

Assume a finite set M of motives. We define the *problem space* $P(M)$ as the set of all chess problems that could be characterized by the motives in M.

Principle 1 (set inclusion)

For a chess problem p let $F(p)$ be the set of motives that a subject must recognize in order to find a correct solution for p. Assume that the difficulty of a chess problem p depends only on $F(p)$, all other factors are taken to be constant for all

[1] These principles are also reported in Albert, Schrepp and Held (1994). Principle 1 is also discussed in Albert and Held (1994).

problems under investigation. To establish a surmise relation on $P(M)$, we can therefore identify a problem $p \in P(M)$ with the set of motives $F(p) \subseteq M$. The relation $\preceq_1$ is defined for all problems $p, q \in P(M)$ by the following condition,

$$p \preceq_1 q :\Leftrightarrow F(p) \subseteq F(q).$$

This means that if a person is able to solve problem q, then she or he is also able to solve any problem p that can be solved with the knowledge of a subset of $F(q)$. From set theory it is clear that $\preceq_1$ is a quasi–order[2] and hence a surmise relation.

It is clear from chess experience that some motives are easier to recognize than others. Therefore, the motives themselves can be ordered with respect to their difficulty. We assume that the differences in the difficulty of motives can be described by a quasi-order $\sqsubseteq_M$ on M. For motives $m_i, m_j \in M$ the interpretation of $m_i \sqsubseteq_M m_j$ is "every person who is able to recognize m_i is also able to recognize m_j".

Because we want to consider also such differences in the difficulty to recognize motives, we have to generalize the set inclusion principle with respect to the relation $\sqsubseteq_M$ on M. We define a relation $\preceq_1'$ for all $p, q \in P(M)$ by,

$$p \preceq_1' q :\Leftrightarrow F(p) \subseteq' F(q),$$

where the relation $\subseteq'$ is for two subsets $F(p), F(q)$ of M characterized through the following definition.

DEFINITION 3.1 Let $\{m_1, \dots, m_k\}, \{m_1', \dots, m_l'\} \subseteq M$. Then $\{m_1, \dots, m_k\} \subseteq' \{m_1', \dots, m_l'\}$, if and only if there exists an injective[3] mapping $f : \{1, \dots, k\} \to \{1, \dots, l\}$ with $m_i \sqsubseteq_M m'_{f(i)}$ for all $i \in \{1, \dots, k\}$. □

The reflexivity of $\subseteq'$ is obvious (chose f as identity). The transitivity of $\subseteq'$ follows because of the transitivity of $\sqsubseteq_M$ and the fact that the convolution of two injective mappings is also injective. Hence $\subseteq'$ is a quasi–order on the power set of M. $\subseteq'$ depends strongly on the relation $\sqsubseteq_M$ on M. This dependency is illustrated by the following example.

EXAMPLE 3.1 Let $M := \{a, b, c\}$. Fig. 2 shows the resulting quasi–order $\subseteq'$ on the power set of M for two different assumptions concerning $\sqsubseteq_M$. In one $a \sqsubseteq_M b \sqsubseteq_M c$ is assumed (all edges), in the other this assumption is dropped (solid edges only). □

Note that for the special case $m_i \sqsubseteq_M m_j \Leftrightarrow i = j$ the relations $\subseteq$ and $\subseteq'$, and therefore also the surmise relations $\preceq_1$ and $\preceq_1'$, are identical. In general, we have $p \preceq_1 q \Rightarrow p \preceq_1' q$, i.e. $\preceq_1$ is included in $\preceq_1'$.

[2] $\preceq_1$ is not antisymmetric, i.e. not a partial order, since different problems may be represented by the same motive set.

[3] A mapping $f : \{1, \dots, k\} \to \{1, \dots, l\}$ is called injective if it fulfills the condition $i \neq j \Rightarrow f(i) \neq f(j)$ for all $i, j \in \{1, \dots, k\}$.

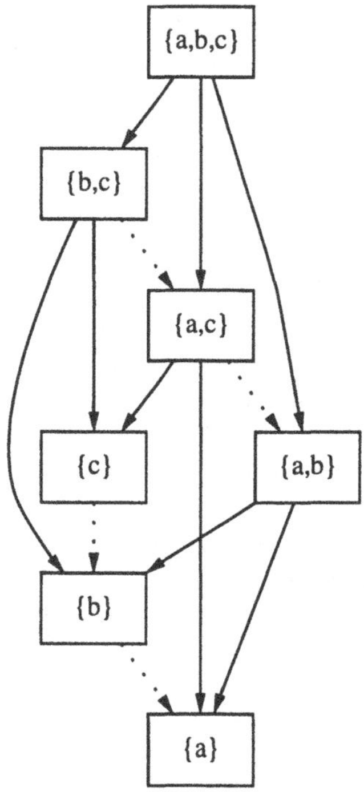

FIG. 2. Surmise relation on $P(M)$, according to "Principle 1" (set inclusion). Solid edges indicate the resulting quasi-order $\sqsubseteq'$, if it is assumed that $x \sqsubseteq_M y \Leftrightarrow x = y$, with $x, y \in \{a, b, c\}$. Dotted edges indicate the additional pairs, if it is assumed that $a \sqsubseteq_M b \sqsubseteq_M c$.

Principle 2 (multiset inclusion)

Within Principle 1, only the occurrence of motives in the solution of a chess problem is taken into account for the construction of a surmise relation. However, it may be necessary to detect a motive more than once to solve a chess problem. So, it seems to be plausible that the multiplicity of motives occurring in the solution of a chess problem also influences its difficulty. To describe this influence we use the concept of a *multiset.* A multiset is a set with the additional property that elements can occur more than once. We write in the following $[x_1, \ldots, x_n]$ for a multiset containing the elements $x_1, \ldots, x_n$. Note that in contrast to usual sets some of the elements can be identical,so $x_i = x_j$ can occur for some $i, j \in \{1, \ldots, n\}$ with $i \neq j$. As an example for the difference between multisets and sets in the usual sense we have $\{x, x\} = \{x\}$ but $[x, x] \neq [x]$. We can define a binary relation $\subseteq_m$, called *multiset inclusion*, on the multisets by:

$$[m_1, \ldots, m_k] \subseteq_m [m'_1, \ldots, m'_l],$$

if and only if there exists an injective function $f : \{1, \ldots, k\} \rightarrow \{1, \ldots, l\}$ with $m_i = m'_{f(i)}$. Hence, a multiset X is included in a multiset Y, if and only if every element occurs at least as frequently in Y as in X. For example, we have $[x, y] \subseteq_m [x, x, y] \subseteq_m [x, x, y, y]$. It is obvious that $\subseteq_m$ is a partial order.

For a chess problem p let $G(p)$ be the multiset of motives that a subject must recognize in order to find a correct solution for p. Suppose that the difficulty of a chess problem p depends only on the multiset $G(p)$ of motives contained in the solution of p. To establish a surmise relation $\preceq_2$ on $P(M)$ it is therefore sufficient, because we can identify a problem p with $G(p)$, to define for problems $p, q \in P(M)$,

$$p \preceq_2 q :\Leftrightarrow G(p) \subseteq_m G(q).$$

As in Principle 1, we want also to consider differences in the difficulty to recognize motives. So, we have to generalize the multiset inclusion principle with respect to the relation $\sqsubseteq_M$ on the motive set M. We define a relation $\preceq'_2$ for $p, q \in P(M)$ by,

$$p \preceq'_2 q :\Leftrightarrow G(p) \subseteq'_m G(q).$$

The relation $\subseteq'_m$ is given by the following definition. Let $L(M)$ be the set of all multisets containing only elements from M.

DEFINITION 3.2 Let $[m_1, \ldots, m_k], [m'_1, \ldots, m'_l] \in L(M)$. Then $[m_1, \ldots, m_k] \subseteq'_m [m'_1, \ldots, m'_l]$, if and only if there exists an injective function $f : \{1, \ldots, k\} \rightarrow \{1, \ldots, l\}$ with $m_i \sqsubseteq_M m'_{f(i)}$ for all $i \in \{1, \ldots, k\}$. □

The reflexivity and transitivity of $\subseteq'_m$ follows, as in Principle 1, from the transitivity of $\sqsubseteq_M$ and the fact that the convolution of injective functions is also an injective function. Note that $\subseteq'_m$ depends strongly on the assumed quasi-order $\sqsubseteq_M$ on the motive set M. This dependency is illustrated by the following example.

EXAMPLE 3.2 Let $M := \{a, b\}$. Fig. 3 shows the resulting quasi-order $\subseteq'_m$ on the set of all multisets containing at least two elements of M for two different assumptions concerning $\sqsubseteq_M$. In one $a \sqsubseteq_M b$ is assumed (all edges), in the other this assumption is dropped (solid edges only). □

Note that for the special case $m_i \sqsubseteq_M m_j :\Leftrightarrow i = j$ we have $\subseteq'_m = \subseteq_m$, the relation $\subseteq'_m$ is identical with the multiset inclusion $\subseteq_m$, which implies trivially that the surmise relations $\preceq_2$ and $\preceq'_2$ are also in this special case identical. In general, we have $p \preceq_2 q \Rightarrow p \preceq'_2 q$, i.e. $\preceq_2$ is included in $\preceq'_2$.

Principle 3 (sequence inclusion)

In Principles 1 and 2, the order in which the motives occur in the solution of a chess problem does not affect the resulting surmise relation. Ignoring this order

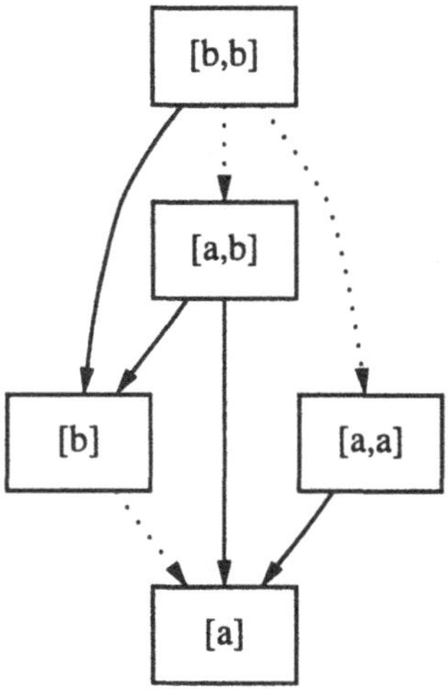

FIG. 3. Surmise relation on $P(M)$, accordingly to "Principle 2" (multiset inclusion). Solid edges indicate the resulting quasi-order $\sqsubseteq'_m$, if it is assumed that $x \sqsubseteq_M y \Leftrightarrow x = y$, with $x, y \in \{a, b\}$. Dotted edges indicate the additional pairs, if it is assumed that $x \sqsubseteq_M y \Leftrightarrow x = y \vee (x = a \wedge y = b)$, with $x, y \in \{a, b\}$.

may be problematical for the analysis of chess problem solving, because of forward and backward search strategies in problem solving. In Principle 3, the order in which the motives occur in the solution of a problem plays a central role in the construction of the surmise relation on $P(M)$. The central idea of Principle 3 is that a problem a is more difficult than a problem b if the sequence of motives that must be recognized by a subject to solve a includes the sequence of motives that must be recognized to solve b.

For a chess problem $p \in P(M)$, let $H(p)$ be the ordered tuple of motives that occur in the solution of the problem. $H(p) = (m_1, \ldots, m_k)$ means that the first motive which occurs is m_1, the second is m_2, etc. Note that one motive can occur more than once within $H(p)$. In this case, the multiplicity of motives is also covered, for example, $m_i = m_j$ for $i \neq j$.

The order in which the motives occur in the solution of a problem[4] is very important for Principle 3. However, it will be clear from our ordering principle that the resulting surmise relation will not change if the order $(m_1, \ldots, m_k)$ is, *for all* problems, inverted to $(m_k, \ldots, m_1)$. We define,

$$M_k := \{(m_1, \ldots, m_k) \mid m_1, \ldots, m_k \in M\},$$

$$M_{\mathbb{N}} := \bigcup_{k \in \mathbb{N}} M_k.$$

[4] The order, in which the motives are recognized by a subject, is not necessarily identical with the order in the solution of a problem, and in the solution plan, respectively.

M_k is the set of all ordered k-tuples of motives from M and $M_{\mathbf{N}}$ is the set of all ordered tuples of motives from M. We define a binary relation on $M_{\mathbf{N}}$ by,

$$(m_1, \ldots, m_k) \sqsubseteq (m'_1, \ldots, m'_l),$$

if and only if there exists a strictly increasing[5] function $f: \{1, \ldots, k\} \to \{1, \ldots, l\}$ with $m_i = m'_{f(i)}$. Hence $(m_1, \ldots, m_k) \sqsubseteq (m'_1, \ldots, m'_l)$, if and only if $(m_1, \ldots, m_k)$ can be obtained by deleting motives from $(m'_1, \ldots, m'_l)$. For example, we have $(x, z) \sqsubseteq (x, y, z)$ but $(x, z) \not\sqsubseteq (z, y, x)$ because we can get (x, z) by deleting y from (x, y, z) but no deletion of an element of (z, y, x) can yield (x, z).

Suppose that under suitable conditions the difficulty of a chess problem $p \in P(M)$ depends only on $H(p)$. To establish a surmise relation $\preceq_3$ on $P(M)$, we identify a problem p with $H(p)$. It is then sufficient to define for problems $p, q \in P(M)$,

$$p \preceq_3 q :\Leftrightarrow H(p) \sqsubseteq H(q).$$

We extend the sequence inclusion principle in order to consider also differences in the difficulty to recognize motives. We define a relation $\preceq'_3$ for $p, q \in P(M)$ by

$$p \preceq'_3 q :\Leftrightarrow H(p) \sqsubseteq' H(q),$$

where the relation $\sqsubseteq'$ is given through the definition below.

DEFINITION 3.3 Let $(m_1, \ldots, m_k), (m'_1, \ldots, m'_l) \in M_{\mathbf{N}}$. Then $(m_1, \ldots, m_k) \sqsubseteq' (m'_1, \ldots, m'_l)$, if and only if there exists a function $f : \{1, \ldots, k\} \to \{1, \ldots, l\}$ that fulfills the following conditions:

1. $\forall i, j \in \{1, \ldots, k\} \quad (i < j \to f(i) < f(j))$,
2. $\forall j \in \{1, \ldots, k\} \quad (m_j \sqsubseteq_M m'_{f(j)})$. □

The reflexivity of $\sqsubseteq'$ is obvious (choose the identity for f). The transitivity of $\sqsubseteq'$ follows because of the transitivity of $\sqsubseteq_M$ and the fact that the convolution of two monotonic increasing functions also has this property. Hence, $\sqsubseteq'$ is a quasi-order, that depends on the relation $\sqsubseteq_M$ on M. The following example illustrates this dependency.

EXAMPLE 3.3 Let $M := \{a, b\}$. Fig. 4 shows the resulting quasi-order $\sqsubseteq'$ on $M_1 \cup M_2 \subseteq P(M)$ for two different assumptions concerning $\sqsubseteq_M$. In one $a \sqsubseteq_M b$ is assumed (all edges), in the other this assumption is dropped (solid edges only). □

For the special case $m_i \sqsubseteq_M m_j :\Leftrightarrow i = j$ we have $\sqsubseteq = \sqsubseteq'$ and therefore $\preceq_3 = \preceq'_3$. In general $p \preceq_3 q \Rightarrow p \preceq'_3 q$, so $\preceq_3$ is included in $\preceq'_3$.

[5] A function $f : \{1, \ldots, k\} \to \{1, \ldots, l\}$ with $i < j \Rightarrow f(i) < f(j)$ is called strictly increasing.

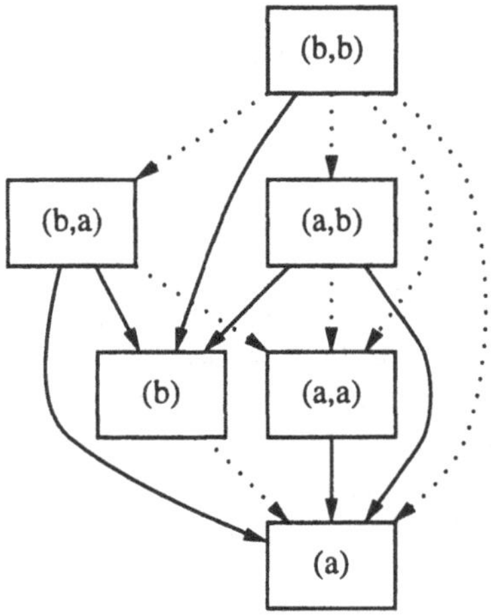

FIG. 4. Surmise relation on $P(M)$, accordingly to "Principle 3" (sequence inclusion). Solid edges indicate the resulting quasi-order $\sqsubseteq$, if it is assumed that $x \sqsubseteq_M y \Leftrightarrow x = y$, with $x, y \in \{a, b\}$. Dotted edges indicate the additional pairs, if it is assumed that $x \sqsubseteq_M y \Leftrightarrow x = y \vee (x = a \wedge y = b)$, with $x, y \in \{a, b\}$.

Comparison of the Principles

A comparison of the surmise relations $\preceq'_1$, $\preceq'_2$, and $\preceq'_3$ resulting from our three ordering principles shows that $\preceq'_1$ is stricter[6] than $\preceq'_2$ and that $\preceq'_2$ is stricter than $\preceq'_3$. We summarize and prove this dependency in the following theorem.

THEOREM: Let M be a set of motives and $p, q \in P(M)$. Then we have

1. $p \preceq'_2 q \Rightarrow p \preceq'_1 q$
2. $p \preceq'_3 q \Rightarrow p \preceq'_2 q$

Proof: First, we show the implication $p \preceq'_2 q \Rightarrow p \preceq'_1 q$. Let $G(p) = [m_1, \ldots, m_k]$ and $G(q) = [m'_1, \ldots, m'_l]$ be the multisets of motives that must be recognized to solve the problems p and q. $p \preceq'_2 q$ implies by definition the existence of an injective function $f : \{1, \ldots, k\} \to \{1, \ldots, l\}$ with $m_i \sqsubseteq_M m'_{f(i)}$. Because we have $F(p) = \{m_1, \ldots, m_k\}$ and $F(q) = \{m'_1, \ldots, m'_l\}$ this trivially implies $F(p) \subseteq' F(q)$. Hence, the implication $p \preceq'_2 q \Rightarrow p \preceq'_1 q$ follows directly.
Second, we show $p \preceq'_3 q \Rightarrow p \preceq'_2 q$. Let $H(p) = (m_1, \ldots, m_k)$ and $H(q) = (m'_1, \ldots, m'_l)$ be the sequences of motives occurring in the solutions of the problems p and q. $p \preceq'_3 q$ implies the existence of a function $f : \{1, \ldots, k\} \to \{1, \ldots, l\}$ with the properties $\forall i, j \in \{1, \ldots, k\}\ (i < j \Rightarrow f(i) < f(j))$ and

[6] We call a relation R to be stricter than a relation R' if R' is included in R($R' \subseteq R$). This is equivalent to the fact that for every x and y the implication $xRy \Rightarrow xR'y$ is true.

$\forall j \in \{1, \ldots, k\}$ $(m_j \sqsubseteq_M m_{f(j)})$. Because every strictly increasing function is also injective, this directly implies $p \preceq_2' q$. □

KNOWLEDGE SPACES FOR CHESS PROBLEMS

For an empirical test of the three ordering principles described above, a set of 16 tactical chess problems has been constructed. For the solution of each of these problems only the motives shown in Table 1 have to be detected. Therefore the motive set M is given through $M = \{F, G, E, C, T, S\}$. Each problem contains four motives at the most. The number of moves necessary for the solution of the problems ranges from one to four, although one move does not necessarily represent one motive. In all problems White moves first and has to find the optimal moves in the given position. The optimal moves do not necessarily lead to a mate. Forcing a draw by stalemate or reaching a winning position can also be optimal solutions. A complete list of the 16 problems, their characterization as tuples of motives and the solutions are provided in Table 2.

In the following, we characterize chess problems by the ordered tuples of motives occurring in their solution. From this characterization as motive tuples, the corresponding characterizations as multisets or sets of motives can easily be derived.

Remember that the surmise relations $\preceq_1'$, $\preceq_2'$, and $\preceq_3'$ derived from the principles "set inclusion", "multiset inclusion" and "sequence inclusion" depend to a great extent on the assumed quasi–order $\sqsubseteq_M$ on the motive set. As mentioned already for the special case $x \sqsubseteq_M y \Leftrightarrow x = y$ for all $x, y \in \{F, G, E, T, C, S\}$, where is assumed that none of the motives can be detected more easily than other motives, we have $\preceq_1 = \preceq_1'$, $\preceq_2 = \preceq_2'$, and $\preceq_3 = \preceq_3'$.

For our ordering principles, we distinguish in the following between two types of different hypothetical problem structures. If we assume that none of the motives can be recognized more easily than other motives, $(m_i \sqsubseteq_M m_j \Leftrightarrow i = j)$, the surmise relations $\preceq_1$, $\preceq_2$, and $\preceq_3$ consist of two nonconnected structures. If we assume that the motive fork can be detected more easily than the motives S, E, C, G, T formalized by $m_i \sqsubseteq_M m_j \Leftrightarrow i = j \vee (m_i = F \wedge m_j \in \{S, E, C, G, T\})$, surmise relations $\preceq_1'$, $\preceq_2'$, and $\preceq_3'$ with additional relational dependencies between problems result.

The surmise relations $\preceq_1$ and $\preceq_1'$ on the problem set are depicted in Fig. 5 as a Hasse–Diagram. The surmise relations $\preceq_2$ and $\preceq_2'$ are depicted in Fig. 6, and the surmise relations $\preceq_3$ and $\preceq_3'$ are depicted in Fig. 7.

Note that the relations $\preceq_2$ and $\preceq_3$ differ only in three relational dependencies. For $\preceq_2$ we have $(G, E, S) \preceq_2 (E, G, S)$, $(E, G, S) \preceq_2 (G, E, S)$, and $(G, E, S) \preceq_2 (E, E, G, S)$, whereas for $\preceq_3$, these relational dependencies do not hold, that means, $(G, E, S) \not\preceq_3 (E, G, S)$, $(E, G, S) \not\preceq_3 (G, E, S)$, and $(G, E, S) \not\preceq_3 (E, E, G, S)$. Because of this small difference it seems to be hard to decide empir-

TABLE 2
Complete List of Chess Problems for Experiment 2

Type	*Position*	*Solution*	*Alternative*
(S)	*White*: Ka1 Qf5 *Black*: Kh8 Pg6	1. Q arbitrary except of Qg6:,Qh5	
(G, S)	*White*: Kh1 Bf2 *Black*: Kc7 Qe5	1. Bg3 Qg3: stale-mate	
(E, G, S)	*White*: Kh1 Bf2 Ktc3 *Black*: Kc7 Qf4 Pc6	1. Ktd5+ cd5: 2. Bg3 Qg3: stale-mate	
(E, E, G, S)	*White*: Ka1 Bc2 Ktf3,d7 *Black*: Kf7 Qc4 Pf6,d6	1. Kte5+ de5:/fe5: 2. Kte5:+ fe5:/de5: 3. Bb3 Qb3: stale-mate	
(C, S)	*White*: Ka3 Pb2 *Black*: Kc5 Rh2 Pa4,b5,c4	1. b4+ ab3:ep/cb3:ep stale-mate	
(G, C, S)	*White*: Kh3 Bg4 Pg2 *Black*: Kg6 Rc2 Ph4,g5,f4	1. Bf5+ Kf5: 2. g4+ hg3:ep/fg3:ep stale-mate	
(T, S)	*White*: Kh5 Pf7 *Black*: Kh7	1. f8R win	
(G, E, S)	*White*: Ka1 Bc2 Kte2 *Black*: Ke6 Qc4 Pe5	1. Bb3 Qb3: 2. Ktd4+ ed4: stale-mate	
(F)	*White*: Kb2 Ktf3 *Black*: Kf7 Qc6	1. Kte5+ draw	
(G, F)	*White*: Kh1 Be3 Ktc3 *Black*: Kc7 Qe5	1. Bf4 Qf4: 2. Ktd5+ draw	
(G, F, F)	*White*: Kg1 Bc6 Kte4,g4 *Black*: Kg8 Qe6 Pg7	1. Bd5 Qd5: 2. Ktf6+ gf6: 3. Ktf6:+ draw	
(G, G, F, F)	*White*: Kb1 Bg3 Kth4,e7 Pf7,c2 *Black*: Kh8 Qf6 Bg7 Ph7	1. f8Q(R)+ Bf8: 2. Be5 Qe5: 3. Ktg6+ hg6: 4. Ktg6:+ draw	1. Ktg6+ hg6: 2.f8Q+ Bf8: 3. Be5 Qe5: 4. Ktg6:+ draw
(G, G, F)	*White*: Kb1 Bg3 Kth4 Pf7,e2 *Black*: Kh8 Qf6 Bg7	1. f8Q(R)+ Bf8: 2. Be5 Qe5: 3. Ktg6+ draw	
(F, F)	*White*: Kb2 Kte3,c3 *Black*: Kc7 Qf4 Pe6	1. Ktd5+ ed5: 2. Ktd5:+ draw	
(T, F)	*White*: Kc1 Pf7,b2 *Black*: Kh7 Qd7	1. f8Kt+ win	
(T, F, F)	*White*: Kb1 Kte4 Pf7 *Black*: Kh7 Qd7 Bg7	1. f8Kt+ Bf8: 2. Ktf6+ draw	1.Ktf6+ Bf6: 2.f8Kt+ draw

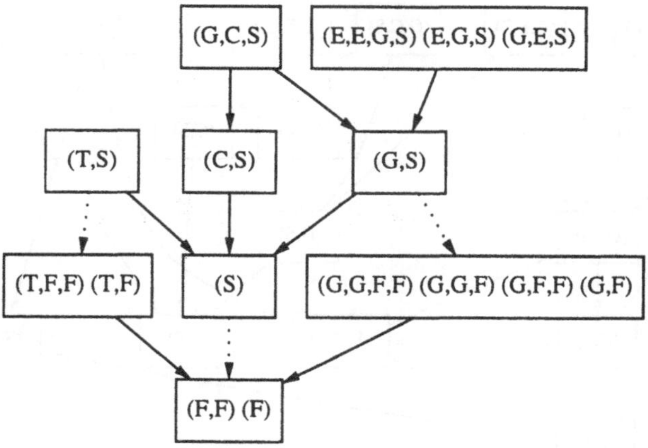

FIG. 5. Solid edges indicate the resulting surmise relation $\preceq_1$ (set inclusion), assuming that $x \sqsubseteq_M y \Leftrightarrow x = y$, with $x, y \in \{F, G, E, T, C, S\}$. All edges indicate the surmise relation $\preceq_1'$, assuming that $x \sqsubseteq_M y \Leftrightarrow x = y \vee (x = F \wedge y \in \{G, E, T, C, S\})$.

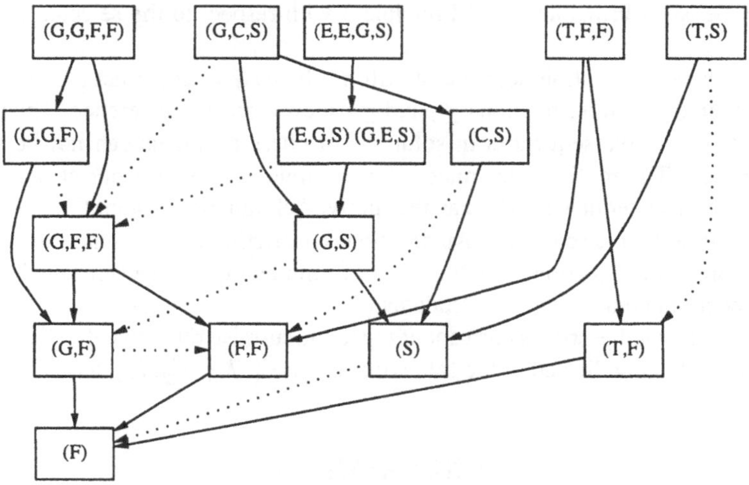

FIG. 6. Solid edges indicate the resulting surmise relation $\preceq_2$ (multiset inclusion). It is assumed that $x \sqsubseteq_M y \Leftrightarrow x = y$, with $x, y \in \{F, G, E, T, C, S\}$. All edges indicate the surmise relation $\preceq_2'$. Here it is assumed that $x \sqsubseteq_M y \Leftrightarrow x = y \vee (x = F \wedge y \in \{G, E, T, C, S\})$.

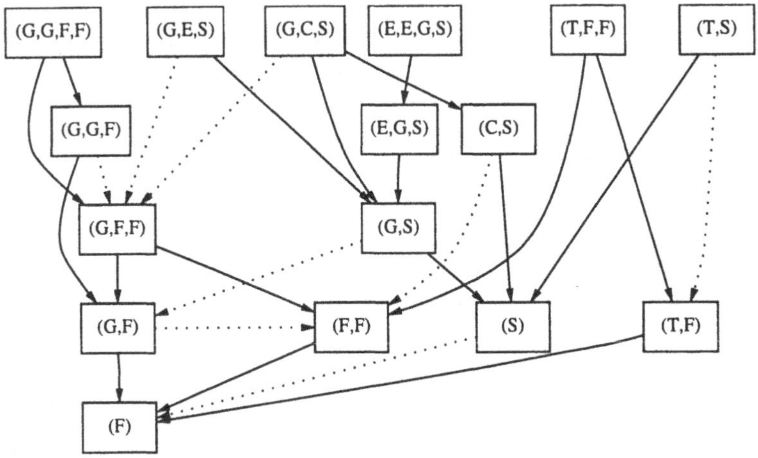

FIG. 7. Solid edges indicate the resulting surmise relation $\preceq_3$ (sequence inclusion). It is assumed that $x \sqsubseteq_M y \Leftrightarrow x = y$, with $x, y \in \{F, G, E, T, C, S\}$. All edges indicate the surmise relation $\preceq_3'$. Here it is assumed that $x \sqsubseteq_M y \Leftrightarrow x = y \vee (x = F \wedge y \in \{G, E, T, C, S\})$.

ically between Principle 2 and Principle 3 with respect to the selected problem set.

We have to mention here that it will not be easy to construct a problem set for which the surmise relations $\preceq_2$ and $\preceq_3$ show a greater difference. The reason is that the motive sequences in solutions of chess problems can not be varied arbitrarily. The motive "stalemate", for example, can only occur at the end of such a motive sequence, whereas the motive "elimination" cannot occur as last motive in a motive sequence. An additional restriction is that some motives can occur only in connection with others. For example, the motive "elimination" can only occur in connection with "stalemate".

The knowledge spaces corresponding to the surmise relations $\preceq_1$,$\preceq_1'$, $\preceq_2$, $\preceq_2'$, $\preceq_3$, and $\preceq_3'$ are indicated in the following as $\mathcal{K}_1$,$\mathcal{K}_1'$, $\mathcal{K}_2$, $\mathcal{K}_2'$, $\mathcal{K}_3$, and $\mathcal{K}_3'$.

EXPERIMENT I

To test the validity of our ordering principles empirically[7], we compare the quasi–ordinal knowledge spaces $\mathcal{K}_1$,$\mathcal{K}_1'$, $\mathcal{K}_2$, $\mathcal{K}_2'$, $\mathcal{K}_3$, and $\mathcal{K}_3'$ derived from these ordering principles with the response patterns of 46 subjects, that worked on the 16 problems described in Table 2. The investigation was conducted at the Psycho-

[7] The data of Experiment I and their analysis concerning "Principle 1" and "Principle 3" are already published in Albert, Schrepp and Held (1994).

logical Institute of the University of Heidelberg.

Method

The experiment was conducted with 37 male and 9 female subjects. Their ages ranged from 14 to 71 years. All of them were familiar with the basic rules of chess. The subjects were recruited through an announcement in the local newspaper. For taking part in the investigation each subject was paid DM 12,–. The experiment was conducted in an experimental room. The problems were presented on a computer screen (SUN–3 workstation with monochrome monitor) and the subjects were required to make their moves using a mouse.

A run of the experiment consists of two main phases: a training phase and an experimental phase. The training phase allows the subjects to learn and practise moving the pieces on the computer chess board. On completion of the training phase, the experimental phase, starts with three very simple "dummy problems", that are not relevant for further analysis. After the presentation of these initial problems was finished, the 16 problems described in the last section were presented in random order.

The subject had 90 seconds to work out a solution plan to a problem. After that, each move had to be made within 30 seconds. If this time is exceeded, a message is displayed requesting the subjects to draw faster. The time limitations were chosen to preclude two kinds of strategy on the part of the subjects: (1) to prevent any subject moving too soon without having thought sufficiently about the solution to the problem, and (2) to prevent the subject continuing to work out a solution while making his or her moves.

After each move by the subject, the computer either made a suitable reply (if the subject's move was correct), or the game was terminated (if either the problem was completed or the subject's move was wrong). At no time do subjects receive feedback concerning the correctness of their answers.

Results

An overview about individual results is provided in Table 3.

First, how well do the data fit our knowledge spaces? Table 4 provides the number of states (elements of the corresponding knowledge space), the number of subjects with response patterns congruent with a state, the number of response patterns not congruent with any state, as well as the number of *different* congruent and not congruent response patterns within the data. Table 5 presents the symmetric distances [8] between response patterns and closest states.

[8] The symmetric distance d between two sets A and B is defined as follows: $d(A, B) :=\mid A \triangle B \mid$, where $A \triangle B = (A \setminus B) \cup (B \setminus A)$.

TABLE 3

Individual Results in Experiment I

Subjects	Distances					Problems															
	K_1	K_2	K_2'	K_3	K_3'	S	GS	EGS	EEGS	CS	GCS	TS	GES	F	GF	GFF	GGFF	GGF	FF	TF	TFF
1,14	1	0	0	0	0	+	+	+	+	+	+	+	+	+	+	+	–	+	+	+	+
2,18,30,37	0	0	0	0	0	+	+	+	+	+	+	+	+	+	+	+	+	+	+	+	+
3	3	0	0	0	0	+	+	–	–	–	–	+	–	+	+	+	–	–	+	+	–
4	4	2	3	1	2	+	+	+	+	+	+	+	–	+	+	–	+	–	+	+	–
5	1	1	1	1	1	+	–	–	–	–	–	+	–	+	–	+	–	–	+	+	+
6	3	2	2	2	2	+	–	–	+	–	–	+	+	+	+	–	–	–	+	+	+
7	3	3	4	3	4	+	+	–	–	–	–	–	–	–	–	–	+	–	+	–	+
8	0	0	1	0	1	+	–	–	–	–	–	+	–	+	–	–	–	–	+	–	–
9,21,45	1	0	0	0	0	–	–	–	–	–	–	–	–	+	–	–	–	–	–	–	–
10	1	0	0	0	0	+	+	+	+	–	–	+	–	+	+	+	+	+	+	+	+
11	1	0	0	0	0	–	–	–	–	–	–	–	–	+	+	–	–	–	+	+	+
12	2	0	0	0	0	+	–	–	–	–	–	+	–	+	+	+	–	–	+	+	+
13,34	1	0	0	0	0	+	–	–	–	+	–	+	–	+	–	–	–	–	+	+	–
15	1	0	0	0	0	+	+	+	+	–	–	+	+	+	+	+	–	+	+	+	+
16	1	1	2	1	2	+	–	–	–	–	–	–	–	–	–	–	–	–	–	+	–
17	2	0	0	0	0	+	+	+	–	+	+	–	+	+	+	+	–	+	+	+	+
19	2	2	2	2	2	–	–	–	–	+	–	–	–	–	–	–	–	–	–	+	–
20	2	0	1	0	1	–	–	–	–	–	–	–	–	+	–	–	–	–	–	+	–
22	0	0	1	0	1	+	–	–	–	–	–	–	–	–	–	–	–	–	–	–	–
23	3	1	1	1	1	+	–	–	+	–	–	–	–	+	+	+	–	–	+	–	–
24	2	1	2	1	2	–	–	–	–	+	–	+	–	+	–	–	–	–	–	–	–
25	0	0	0	0	0	+	–	–	–	–	–	+	–	+	–	–	–	–	+	+	+
26	1	0	0	0	0	–	–	–	–	–	–	–	–	+	–	–	–	–	+	+	–
27,38	1	1	1	1	1	–	–	–	–	+	–	–	–	+	–	–	–	–	+	–	–
28,41	2	1	1	1	1	–	–	–	–	+	–	–	–	+	–	–	–	–	–	–	–
29	2	2	2	2	2	+	–	–	–	+	+	–	–	+	–	–	–	–	–	+	+
31	2	0	1	0	1	+	+	+	+	+	–	–	+	+	+	–	–	–	+	+	–
32	1	0	0	0	0	+	–	–	–	+	–	–	–	+	–	–	–	–	+	+	–
33	2	2	2	1	1	+	+	+	+	–	–	+	–	+	+	–	+	+	+	+	+
35	4	1	2	1	2	+	+	+	–	–	–	+	+	+	+	–	+	–	+	+	–
36,42	1	0	0	0	0	+	–	–	–	–	–	–	–	+	–	–	–	–	+	+	–
39	1	1	1	1	1	–	–	–	–	–	–	–	–	–	–	–	–	–	+	–	–
40	1	0	0	0	0	+	–	–	–	–	–	–	–	+	–	–	–	–	–	–	–
43	2	1	1	1	1	–	–	–	–	+	–	+	–	+	–	–	–	–	+	+	–
44	2	0	1	0	1	+	–	–	–	–	–	–	–	+	–	–	–	–	–	+	–
46	2	1	1	1	1	–	–	–	–	+	–	–	–	+	–	–	–	–	+	+	–

A "+" indicates that the problem was solved by the corresponding subject, while a "–" indicates that this was not the case.

TABLE 4
Possible States and Response Patterns for Experiment I

Knowl. space	*Number of states*	*Number of congruent patterns*	*Number of non congruent patterns*	*Number of different congruent patterns*	*Number of different non congruent patterns*
$\mathcal{K}_1$	85	7	39	4	32
$\mathcal{K}'_1$	35	5	41	2	34
$\mathcal{K}_2$	575	27	19	20	16
$\mathcal{K}'_2$	213	22	24	15	21
$\mathcal{K}_3$	1025	28	18	20	16
$\mathcal{K}'_3$	368	23	23	15	21

TABLE 5
Frequency of Distances Between Response Patterns and Closest States for Experiment I

Knowl. space	*Distances*					*Average Distance*
	0	*1*	*2*	*3*	*4*	
$\mathcal{K}_1$	7	20	13	4	2	1.43
$\mathcal{K}'_1$	5	21	12	4	4	1.59
$\mathcal{K}_2$	27	13	5	1	0	0.57
$\mathcal{K}'_2$	22	15	7	1	1	0.78
$\mathcal{K}_3$	28	14	3	1	0	0.50
$\mathcal{K}'_3$	23	15	7	0	1	0.72

Number of subjects who solved all problems: 4
Number of subjects who solved no problem: 0

Discussion

The results of the experiment show that the principle "set inclusion" seems to be less adequate for the ordering of chess problems concerning their difficulty than the principles "multiset inclusion" and "sequence inclusion". As expected, it seems to be impossible to decide on the basis of this results between "multiset inclusion" and "sequence inclusion".

Despite the fact that the quasi-ordinal knowledge spaces $\mathcal{K}_1$,$\mathcal{K}'_1$, $\mathcal{K}_2$, $\mathcal{K}'_2$, $\mathcal{K}_3$, and $\mathcal{K}'_3$ are assumed to be purely deterministic, in comparing them with response patterns we have to consider influences like "lucky guesses" or "careless errors" (see Falmagne & Doignon, 1988) [9].

[9] The probability of lucky guesses and careless errors must be very low. If we assume for instance

Table 5 shows that for the quasi-ordinal knowledge spaces derived from Principles 2 and 3 the symmetric distances between response patterns and closest states are rather small. So most of the deviations between response patterns and the hypothetical knowledge spaces derived from these principles may be explained by such influences like lucky guesses or careless errors. Therefore, these principles seem to be adequate for our set of chess problems.

In analyzing the results, special attention should be paid to the proportions between the number of states and the number of nonstates (i.e. the subsets of the problem set that do not belong to the corresponding knowledge space), on the one hand, and the proportions between the number of response patterns congruent with states and the number of response patterns not congruent with states, on the other. The best fitting space $\mathcal{K}_3$ contains only 1.6 % of the theoretically possible response patterns, while it contains 61 % of the observed response patterns.

The additional assumption that the motive "fork" can be recognized more easily than all of the other motives leads for both principles, "multiset inclusion" and "sequence inclusion" to a reduction in the number of hypothetical knowledge states. It therefore comes as no surprise that this assumption leads to a larger number of response patterns which do not agree with the quasi-ordinal knowledge spaces derived from these principles. Most of them, however, have a small symmetric distance.

In comparing our ordering principles with respect to their adequacy to describe the difficulty of chess problems we have to deal with the problem that the quasi-ordinal knowledge spaces $\mathcal{K}_1$,$\mathcal{K}_1'$, $\mathcal{K}_2$, $\mathcal{K}_2'$, $\mathcal{K}_3$, and $\mathcal{K}_3'$ differ extremely in size. Because both the average symmetric distance and the number of congruent response patterns depend on the size of a knowledge space, these numbers can not be interpreted directly. An approach to solve this problem is presented in the general discussion.

EXPERIMENT II

In a replication of our first experiment, which was conducted at the University of Graz (Austria), the response patterns of 46 subjects were compared with the quasi-ordinal knowledge spaces derived from our ordering principles.

Method

The experiment was conducted with 38 male and 8 female subjects. All of them were familiar with the basic rules of chess. 21 subjects were members of chess clubs, whereas the rest were pupils or students of the University of Graz, and were

that the probability of both careless errors and lucky guesses is 0.05, then the probability that the response pattern of the subject is equal to the subject's knowledge state is $0.95^{16} \approx 0.44$.

pure hobby-players. Therefore, the chess-playing ability of the subjects varied over a wide range. The subjects were not paid for taking part in the investigation. The ages of the subjects ranged from 6 to 64 years. The experiment was conducted partly in a chess club, a school, and in a chess cafe. The problems were presented on a laptop-computer. The laptop was operated by the experimenter, so the subjects made their moves verbally and the experimenter entered these moves into the computer.

The experiment consists, similar to Experiment I, of two phases, a instructional phase and an experimental phase. In the instructional phase the experimental setting and the way the subjects can give their solutions were explained. After the instructional phase was finished the experimental phase started with three simple "dummy problems" (identical with the dummy problems used in Experiment I), which are not relevant for further analysis. Then, the 16 problems described in Table 2 were presented in random order.

The time limitations are similar to Experiment I, the subject had 90 seconds to work out a solution plan and, after that, each move had to be made within 30 seconds. If a subject exceeds the time limitations, the experimenter requires the subject verbally to make a move. After each move of the subject, a suitable reply was shown on the screen (if the subject's move was correct), or the game was terminated (if either the problem was completed or the subjects move was wrong). At no time do the subjects receive feedback concerning the correctness of their answers.

Results

An overview about individual results is given in Table 6.

In Table 7, the number of states, the numbers of congruent and noncongruent response patterns and the number of different congruent and noncongruent response patterns is shown. Table 8 provides an overview about the observed symmetric distances between response patterns and closest states.

Discussion

A comparison of these results with the results of Experiment I show that they are nearly identical. The differences in the experimental setting had almost no influence on the results. Another remarkable point is that in Experiment II, 21 of the 46 subjects were members of chess clubs, whereas in Experiment I almost all subjects were pure hobby players. This implies that the characterization of the problems' difficulty, which is given through our surmise relations, is valid over a wide range of chess playing ability.

As in Experiment I, the principle "set inclusion" seems to be less adequate for the ordering of chess problems than the principles "multiset inclusion" and

Subjects	*Distances* $\mathcal{K}_1$	$\mathcal{K}_2$	$\mathcal{K}_2'$	$\mathcal{K}_3$	$\mathcal{K}_3'$	*Problems* S	GS	EGS	EEGS	CS	GCS	TS	GES	F	GF	GFF	GGFF	GGF	FF	TF	TFF
1	2	1	1	1	1	–	–	–	–	–	–	–	+	+	–	–	–	–	+	+	–
2	2	0	0	0	0	+	+	–	–	–	–	–	–	+	+	+	–	–	+	–	–
3,32	1	0	0	0	0	+	–	–	–	–	–	–	–	+	–	–	–	–	–	–	–
4,36	1	0	0	0	0	–	–	–	–	–	–	–	–	+	–	–	–	–	–	–	–
5	1	1	1	1	1	+	+	+	+	–	–	+	+	+	+	+	+	+	–	+	+
6,16,25,29	0	0	0	0	0	+	+	+	+	+	+	+	+	+	+	+	+	+	+	+	+
7	3	0	0	0	0	+	+	+	–	–	–	+	+	+	+	+	–	–	+	+	+
8	3	1	1	1	1	+	+	–	+	–	–	+	+	+	+	+	–	–	+	+	+
9	2	1	1	1	1	+	–	–	–	–	–	–	–	+	–	+	–	–	–	–	–
10	1	0	0	0	0	–	–	–	–	–	–	–	–	+	–	–	–	–	–	–	–
11	2	1	3	1	3	+	–	–	–	+	–	+	–	+	–	–	–	+	–	–	–
12	2	2	2	2	2	+	+	+	+	–	+	+	+	+	+	+	+	–	+	+	+
13	2	1	2	0	2	+	+	–	–	–	–	+	+	+	–	–	–	–	+	+	–
14	2	1	1	0	1	+	+	–	–	–	–	+	+	+	+	–	–	–	+	+	+
15	1	1	1	1	1	+	+	+	+	–	–	+	+	+	+	+	+	–	+	+	+
17	1	0	0	0	0	–	–	–	–	+	–	–	–	+	–	–	–	–	–	–	–
18	2	0	1	0	1	+	–	–	–	–	–	+	–	+	–	–	–	–	–	+	–
19	0	0	0	0	0	+	–	–	–	–	–	–	–	–	–	–	–	–	–	–	–
20	0	0	0	0	0	+	–	–	–	+	–	+	–	+	–	–	–	–	+	+	+
21	2	1	1	1	1	+	+	+	+	–	+	+	+	+	+	+	–	+	+	+	+
22	3	0	0	0	0	+	+	+	–	+	+	–	+	+	+	+	–	–	+	+	+
23	4	4	4	3	3	+	+	–	–	–	+	+	+	+	+	+	+	–	+	–	+
24	2	1	1	1	1	+	+	+	–	–	–	+	+	+	+	+	+	–	+	+	+
26	2	2	2	2	2	+	–	–	–	–	–	–	+	+	+	+	+	–	+	+	+
27	1	0	0	0	0	+	+	+	–	–	–	+	+	+	+	+	+	+	+	+	+
28	1	1	1	1	1	+	+	+	+	–	+	+	+	+	+	+	+	+	+	+	+
30	3	1	2	1	2	+	–	–	–	–	+	–	–	+	–	–	–	–	–	+	–
31	3	1	2	1	2	–	–	–	–	+	–	–	–	+	–	–	–	–	–	+	–
33	1	1	1	1	1	–	–	–	–	–	–	–	–	+	–	–	–	–	+	–	+
34	1	0	0	0	0	+	–	–	–	–	–	–	–	+	–	–	–	–	+	+	–
35	3	3	3	2	3	+	–	–	–	+	+	+	+	+	–	–	–	–	–	+	+
37	1	0	0	0	0	+	–	–	–	–	–	+	–	+	–	–	–	–	+	+	–
38	1	1	2	1	2	+	+	–	–	–	–	–	–	+	–	–	–	–	+	–	+
39	2	1	1	1	1	+	–	–	–	+	–	+	+	+	–	–	–	–	+	+	–
40	1	0	0	0	0	+	–	–	–	–	–	+	–	+	+	+	–	+	+	+	+
41	1	1	1	1	1	–	–	–	–	–	–	+	–	+	–	–	–	–	+	+	+
42	3	1	2	1	2	+	–	–	–	–	–	+	+	+	–	–	–	–	–	+	–
43	1	0	0	0	0	–	–	–	–	–	–	–	–	+	–	–	–	–	+	+	–
44	0	0	0	0	0	+	–	–	–	–	–	–	–	+	–	–	–	–	+	–	–
45	1	0	0	0	0	+	+	–	–	–	–	–	–	+	+	–	–	–	+	–	–
46	1	0	1	0	1	+	–	–	–	–	–	+	–	+	+	–	–	–	+	–	–

A "+" indicates that the problem was solved by the corresponding subject, while a "–" indicates that this was not the case.

TABLE 7
Possible States and Response Patterns for Experiment II

Knowl. space	*Number of states*	*Number of congruent patterns*	*Number of non congruent patterns*	*Number of different congruent patterns*	*Number of different non congruent patterns*
$\mathcal{K}_1$	85	7	39	4	36
$\mathcal{K}'_1$	35	6	40	3	37
$\mathcal{K}_2$	575	23	23	18	22
$\mathcal{K}'_2$	213	20	26	16	24
$\mathcal{K}_3$	1025	25	21	20	20
$\mathcal{K}'_3$	368	20	26	16	24

TABLE 8
Frequency of Distances Between Response Patterns and Closest States for Experiment II

Knowl.			*Distances*			*Average*
space	*0*	*1*	*2*	*3*	*4*	*Distance*
$\mathcal{K}_1$	7	18	13	7	1	1.50
$\mathcal{K}'_1$	6	15	14	10	1	1.67
$\mathcal{K}_2$	23	19	2	1	1	0.65
$\mathcal{K}'_2$	20	16	7	2	1	0.87
$\mathcal{K}_3$	25	17	3	1	0	0.57
$\mathcal{K}'_3$	20	16	7	3	0	0.85

Number of subjects who solved all problems: 4
Number of subjects who solved no problem: 0

"sequence inclusion", whereas it seems to be impossible to decide on the basis of these results between the last two principles.

Examine in Table 8 the observed symmetric distances between response patterns and closest states and compare the number of states with the number of response patterns congruent with states and response patterns not congruent with states. This examination shows that the principles "multiset inclusion" and "sequence inclusion" seems to be adequate for the ordering of tactical chess problems.

The additional assumption concerning the motive "fork" leads, as in Experiment I, to a reduction of congruent response patterns and to an increase in the average symmetric distances. Because the quasi-ordinal knowledge spaces $\mathcal{K}'_2$ and $\mathcal{K}'_3$ derived from this assumption are contained in the corresponding spaces, $\mathcal{K}_2$ and $\mathcal{K}_3$ we cannot conclude from this increase of empirical violations that the additional assumption concerning the motive "fork" is inadequate. We refer to this problem in the general discussion below.

GENERAL DISCUSSION

In the discussion of our experiments we already mentioned that it is impossible to compare the quality of our quasi-ordinal knowledge spaces using only the observed number of congruent patterns and the observed average symmetric distance between response patterns and spaces. The reason is that these numbers depend on the size of a space. For smaller spaces we will expect more violations than for larger spaces. In the following we will suggest three approaches to solve this problem. These approaches try to compare spaces of different size by relativizing the number of observed violations to the size of the space[10].

First, we discuss an approach of van Leeuwe (1974). This approach evaluates the fit of a surmise relation $\preceq$ to a binary data matrix by comparing the observed correlations between items with the expected correlation if it is assumed that $\preceq$ is correct. The fit between surmise relation and data is measured by the so called *correlational agreement coefficient* CA, which is defined by

$$\mathrm{CA} := 1 - \frac{2}{m(m-1)} \sum_{i<j} (r_{ij} - r^*_{ij})^2 \ ,$$

where m is the number of items, r_{ij} is the Pearson–correlation (Phi-coefficient) between item i and item j, and r^*_{ij} is defined by

$$r^*_{ij} = \begin{cases} 1 & \text{if } i \preceq j \wedge j \preceq i \\ \sqrt{(1-p_i)p_j/(1-p_j)p_i} & \text{if } i \preceq j \wedge j \not\preceq i \\ \sqrt{(1-p_j)p_i/(1-p_i)p_j} & \text{if } i \not\preceq j \wedge j \preceq i \\ 0 & \text{otherwise} \end{cases}$$

Here p_i is the proportion of subjects that solved item i and p_j the proportion of subjects that solved item j. The higher the value of CA is, the better is the fit between surmise relation and data. For a detailed description of this approach see van Leeuwe (1974).

A second approach for the comparison of surmise relations concerning their fit to a given data set is to count for each relational dependency $x \preceq y$ in a surmise relation $\preceq$ how often it is violated. A violation means here that a subject solved x and failed in solving y. The fit between a surmise relation $\preceq$ and a binary data

[10] We have to mention here that all three measures are pragmatical approaches to compare the fit of spaces to observed data, which are theoretically not well founded. Models, which relate space size and random influences (lucky guesses or careless errors) to the number of observed violations of a space are, at the moment, not available.

matrix is then measured by the *violational coefficient* VC defined by,

$$\mathrm{VC} := \frac{1}{n(|\preceq| - m)} \sum_{i,j} v_{ij} ,$$

where n denotes the number of subjects, m the number of problems, $|\preceq|$ the number of relational dependencies in $\preceq$ and v_{ij} is the number of violations of the relational dependency $i \preceq j$ if this dependency is contained in $\preceq$ and 0 otherwise. We use $(|\preceq| - m)$ in the denominator of the fraction, because the m relational dependencies of the form $i \preceq i$ can not be violated empirically. VC can be interpreted as the average number of violations of relational dependencies $i \preceq j$ with $(i \neq j)$ contained in $\preceq$. Therefore the lower the value of VC is, the better is the fit of $\preceq$ to the data.

A third approach, described in Schrepp (1993), makes use of the average symmetric distance between a knowledge structure and a binary data matrix. The fit between a knowledge structure $\mathcal{K}$ and the data matrix is measured by the so-called *distance agreement coefficient* DA defined as,

$$\mathrm{DA} := \frac{\mathrm{ddat}}{\mathrm{dpot}} ,$$

where ddat is the observed average minimal symmetric distance between $\mathcal{K}$ and the observed set of response patterns and dpot is the average minimal symmetric distance between $\mathcal{K}$ and the power set of the item set[11]. The value dpot can be interpreted as the expected average symmetric distance if $\mathcal{K}$ contains no informations concerning the solving behavior of subjects, in other words if the data are randomly chosen. The value of dpot decreases with the size of $\mathcal{K}$. Hence, a decrease of ddat with an increase of the size of $\mathcal{K}$ is compensated by an increase of dpot. A lower value of DA indicates a better fit of $\mathcal{K}$ to the data. For more details concerning this approach, see Schrepp (1993).

Table 9 shows for each of our surmise relations $\preceq_1, \preceq_1', \preceq_2, \preceq_2', \preceq_3$, and $\preceq_3'$, respectively the corresponding quasi-ordinal knowledge spaces, the values of the coefficients CA, VC, and DA. Because there are no systematical differences between the data from Experiment I and Experiment II, we have calculated the values of the coefficients[12] for the union of the data sets.

The three measures, although they are based on different ideas of relativizing the number of observed violations to the size of the space, coincide perfectly in the assertion that the principle "set inclusion" is less adequate for the description of the difficulty of chess problems than the principles "multiset inclusion" and

[11] Formally we have ddat $:= \sum_{P \in \mathcal{D}} \min(\{d(P, K) \mid K \in \mathcal{K}\}) / \mid \mathcal{D} \mid$, where $\mathcal{D}$ denotes the set of all observed response patterns and dpot $:= \sum_{P \in \mathcal{P}} \min(\{d(P, K) \mid K \in \mathcal{K}\}) / \mid \mathcal{P} \mid$, where $\mathcal{P}$ denotes the power set of the item set.

[12] Remember that a higher value of CA indicates a better fit of a space to the data, whereas for VC and DA the converse holds.

TABLE 9
Values of the Coefficients CA, VC and DA for Our Data From Experiment I and Experiment II

Knowl. space	*CA*	*VC*	*DA*
$\mathcal{K}_1$	0.831	0.057	0.34
$\mathcal{K}'_1$	0.863	0.054	0.34
$\mathcal{K}_2$	0.842	0.025	0.18
$\mathcal{K}'_2$	0.863	0.027	0.21
$\mathcal{K}_3$	0.836	0.024	0.17
$\mathcal{K}'_3$	0.867	0.027	0.21

"sequence inclusion". On the basis of our results, it is impossible to decide between the last two principles, because the values of all three coefficients are nearly identical for the spaces $\mathcal{K}_2$ and $\mathcal{K}_3$.

Remember that the relations $\preceq'_2$ and $\preceq'_3$ differ only with respect to the relational dependencies $(G,E,S) \preceq'_2 (E,G,S)$, $(E,G,S) \preceq'_2 (G,E,S)$ and $(G,E,S) \preceq'_2 (E,E,G,S)$, which do not hold for $\preceq'_3$.

The relational dependency $(G,E,S) \preceq'_2 (E,G,S)$ is violated by 10 of our 92 subjects: 10 subjects solved (E,G,S) and failed in solving (G,E,S). The relational dependency $(E,G,S) \preceq'_2 (G,E,S)$ is violated by 3 subjects, whereas $(G,E,S) \preceq'_2 (E,E,G,S)$ is violated by 14 subjects. So at least the two relational dependencies $(G,E,S) \preceq'_2 (E,G,S)$, and $(G,E,S) \preceq'_2 (E,E,G,S)$ are not supported by our data.

The spaces $\mathcal{K}'_2$ and $\mathcal{K}'_3$, which are based on the additional assumption that the motive "fork" can be recognized more easily than the other motives in our motive set, show accordingly to the DA and VC coefficients a lower fit to the data than the spaces $\mathcal{K}_2$ and $\mathcal{K}_3$, whereas for the CA coefficient the converse holds. Therefore, no clear assertion concerning the adequacy of this additional assumption can be drawn. The surmise relations $\preceq'_2$ and $\preceq'_3$ based on this additional assumption show a better approximation to the observed correlations between our chess problems than the relations $\preceq_2$ and $\preceq_3$. However, they also show a worse approximation concerning the violation of relational dependencies contained in them and observed minimal symmetric distances relativized to the size of the corresponding spaces.

Another point that should be discussed is the suitability of motives for the characterization of chess problems. Motives may only be adequate as "exclusive" components of simple problems as used in our investigations. However, other features of the problems may influence their difficulty. For example, the difficulty of a problem may also be dependent on the number of pieces that appear on the board. However, a surmise relation which is solely based on the number of pieces, for example, leads to an average symmetric distance of 2.74 in Experiment I and of 2.63 in Experiment II, which are extremely high compared to the average symmetric distances of our best space $\mathcal{K}_3$ (0.5 and 0.57 in Experiment I and Experiment II) respectively. In our experiments, we tried to minimize the influence of other problem elements by striving for maximal homogeneity of the problem set. With more complex problems, the set of relevant problem components will still contain motives, but other factors that are indispensable in the knowledge of a good chess player may become relatively more important. More information concerning these other factors can be gained from the investigations of De Groot (1965). However, it seems possible to capture also these factors by the theory of knowledge spaces or one of its generalizations.

Our investigation was based on an extension of the theory of Doignon and Falmagne (1985, 1998). We have provided principles for the establishment of surmise relations and quasi-ordinal knowledge spaces that are primarily dependent on problem components. In addition to Doignon and Falmagne, who do not consider the properties of the problems themselves, we are particularly interested in those components of problems that may be fundamental to the surmise relation and the corresponding quasi-ordinal knowledge space. In both our theoretical and empirical investigations we were able to show that the theory of knowledge spaces can be used for testing psychological assumptions, which may be fruitful for the theory of problem solving in general.

ACKNOWLEDGEMENTS

The research reported in this paper was supported by Grant Lu 385/1 of the Deutsche Forschungsgemeinschaft to J. Lukas and D. Albert at the University of Heidelberg. We are grateful to M. Koppen for his important remarks and for suggesting 'Principle 2'. We also thank M. Kadijk for his support in writing the experimental software for Experiment I and P. Mörtl and H. Mariacher for conducting Experiment II.

REFERENCES

Airasian, P. W., & Bart, W. M. (1973). Ordering theory: A new and useful measurement model. *Educational Technology, May*, 56–60.

Albert, D., & Held, T. (1994). Establishment of knowledge spaces by a systematical construction of problems. In: D. Albert (Eds.), *Knowledge structures* (pp.81-115). New York: Springer.

Albert, D., Schrepp, M., & Held, T. (1994). Construction of knowledge spaces for problem solving in chess. In: G. Fischer & D. Laming (Eds.), *Contributions to mathematical psychology, psychometrics, and methodology* (pp. 123–135). New York: Springer.

Bart, W. M., & Krus, D. J. (1973). An ordering-theoretic method to determine hierarchies among items. *Educational and Psychological Measurement, 33*, 291–300.

De Groot, A. D. (1965). *Thought and choice in chess.* The Hague: Mouton & Co.

Doignon, J. P. (1994). Knowledge spaces and skill assignments. In: G. Fischer & D. Laming (Eds.), *Contributions to mathematical psychology, psychometrics, and methodology* (pp 111–121). New York: Springer.

Doignon, J. P., & Falmagne, J.-C. (1985). Spaces for the assesment of knowledge. *International Journal of Man-Machine Studies, 23*, 175–196.

Doignon, J.-P., & Falmagne, J.-C. (1998). *Knowledge spaces.* Berlin: Springer.

Dowling, C. E. (1991). *Constructing knowledge structures from the judgement of experts.* Habilitationsschrift Technische Universität, Braunschweig.

Dowling, C. E. (1993). Applying the basis of a knowledge space for controlling the questioning of an expert. *Journal of Mathematical Psychology, 37*, 21–48.

Held, T. (1993). *Establishment and empirical validation of problem structures based on domain specific skills and textual properties—A contribution to the "Theory of Knowledge Spaces".* Unpublished doctoral dissertation, University of Heidelberg, Germany.

Koppen, M. (1993). Extracting human expertise for constructing knowledge spaces: An algorithm. *Journal of Mathematical Psychology, 37*, 1–20.

Koppen, M., & Doignon, J.-P. (1990). How to build a knowledge space by querying an expert. *Journal of Mathematical Psychology, 34*, 311–331.

Korossy, K. (1993). *Modellierung von Wissen als Kompetenz und Performanz.* [Modeling of knowledge as competence and performance.] Unpublished doctoral dissertation, University of Heidelberg, Germany.

Schrepp, M. (1993). *Über die Beziehung zwischen kognitiven Prozessen und Wissensräumen beim Problemlösen* [On the relation between cognitive processes and knowledge spaces in problem solving]. Unpublished doctoral dissertation, University of Heidelberg, Germany.

Van Leeuwe, J. F. J. (1974). Item Tree Analysis. *Nederlands Tijdschrift voor de Psychologie, 29*, 475–484.

4

An Integrated Approach for Constructing, Coding, and Structuring a Body of Word Problems

Theo Held
Martin-Luther-Universität Halle-Wittenberg

In this chapter, it is shown how a body of word problems can be generated on the basis of "problem grammars", how an efficient and unique coding of the problems by "components" can be performed, and how the derivation of problem structures is facilitated through formal principles working on structures of problem components. Furthermore it is demonstrated how the influence of textual properties of word problems on solution performance can be considered in the process of structuring the body of problems.

The domain under consideration is "elementary stochastics." For the reported experimental investigation 18 word problems have been constructed by means of a problem grammar. The problems vary—aside from domain-specific demands—with respect to the textual properties of "coherence" and "cohesion." Based on problem coding, alternative problem structures have been established that are tested in a fully computerized investigation with $N = 50$ subjects. The results show that the data can be explained very well by the problem structure which is based on a "demand multiassignment" concerning the textual problem component.

INTRODUCTION

The central question of numerous investigations in the area of knowledge space theory—and also of this chapter—is how structures can be imposed on a knowledge domain. In general, there are three main directions that can be followed. First, information can be gained by consulting experts. Algorithms for systematic and economical acquisition of an expert's expertise have, among others, been developed by Dowling (1993), Kambouri, Koppen, Villano, and Falmagne (1994) and Koppen (1993). Second, one can attempt to derive the structure of a domain from existing data (i. e., response patterns). This can be done in order to establish deterministic structures (cf. Bart & Krus, 1973; van Buggenhaut & Degreef, 1987; van Leeuwe, 1990) or probabilistic structures (cf. Falmagne, 1989; Villano, 1991). The third direction is concerned with the analysis of tasks being associated with solving problems in a domain. The analysis can be based on domain-specific (e. g., mathematical) facts, psychological models and theories, or pedagogical factors (e. g., curricular, didactic). The approaches developed (cf. Albert & Held, 1994; Held, 1993; Lukas, 1991) provide principles for deriving structures on a domain from problems or representations of problems.

Such principles are the main subject matter of the work presented in this chapter. It shows how constructive methods can be used for generating sets of problems related to a domain. Problem generation is organized in a way that allows the definition of variable problem elements to which formal principles for structuring problem sets can be applied. It is stated that a variation of well-defined elements of problems entails a variation of "problem demands." The postulation of relationships between problem elements and demands is based on domain-specific and psychological assumptions.

In the following section, a method for problem generation based on "problem grammars" is introduced. The way of generating and defining problems provides the basis for defining specific elements of problems (problem coding). Then, principles for building problem structures that directly work on problem codings, are discussed. Because the application of principles for structuring requires presuppositions about demands associated with problem elements, the concept of "demand assignments" is elaborated.

In the last section, an experimental investigation that makes use of the outlined theoretical results is described. The problem demands relevant to the investigation are supposed to be of mathematical and textual nature. In order to characterize textual demands, a few of annotations about determining textual properties (coherence and cohesion) of the word problems is made.

PROBLEM CONSTRUCTION

In this work, a clear distinction between problems (i. e., problem texts) and problem codings is proposed. Problem generation is concerned with problem texts, whereas problem coding depends both on the generated texts and on considerations about the effect that problem elements have on problem difficulty. In general there are, on the one hand, principles for generating problem texts and, on the other hand, principles for generating problem codings.

This section is concerned with demonstrating how sets of word problems can be generated on the basis of systems of *formal rules*. Because word problems consist of sequences of words, numbers, and mathematical symbols, we can make use of the well developed theoretical framework of *"generative grammars"* in order to generate problem texts and problem sets (Held, 1992, 1993). Generally, grammars consist of a finite set of basic elements and a set of operations that are defined with respect to a combination of the elements. The definition of grammars is not limited to natural languages: In principle, every set of objects that is structured on the basis of *rules* can be described by a grammar. A precise formulation of a theory of *formal* grammars is given by Chomsky (1959). "Phrase structure grammars" are probably the most familiar type of formal grammars. In the case of the generation of word problems, however, there is no given language that is to be described by a grammar. Here a (problem)-language is *defined* by means of the grammar. First, consider two general definitions:

DEFINITION 4.1 (Chomsky, 1959) A *language* L is a collection of sentences of finite length, all constructed from a finite alphabet of symbols. □

According to Chomsky (1959) a grammar of a language L is a function whose range is exactly L. This means that by the application of the grammar, all elements of the language (the sentences) can be generated.

DEFINITION 4.2 A *phrase structure grammar* G consists of a quadruple

$$G = (V_T, V_N, R, S),$$

with V_T a set of *terminal symbols* (the "terminal vocabulary"), V_N a set of *non terminal symbols* (the "non terminal vocabulary"), $S \in V_N$ a special *initial symbol* (the "sentence" S). A sequence formed by concatenating elements of a vocabulary V is called a *string in* V. R is a finite set of *rewriting rules* of the form $\varphi_i \rightarrow \psi_i$ $(i = 1, 2, \ldots, r)$, where φ_i and ψ_i are strings of symbols. □

A sentence is generated by applying the rewriting rules on strings of symbols (starting with '$S \rightarrow \ldots$'). If a string is obtained that completely consists of terminal symbols, the generation procedure is finished. This "transformation" of S into a string φ_n is called a *derivation* of φ_n. A detailed formal description of grammatical problem construction is given in Held (1993). In this chapter the principles of problem construction are demonstrated by an illustrative example. Consider the following word problem:

(1) A grocer bought 17 dozen pears for $ 4.65.
(2) Five dozen of the pears spoiled.
(3) The grocer wants to make a profit equal to $\frac{3}{5}$ of the total cost.
(4) At what price per dozen must he sell the remaining pears?

Generally, this simple arithmetical problem consists of two main parts. The first part—sentences (1) to (3)—provides a few facts that are prerequisites for the solution. The second part is the question of sentence (4), which asks for a value being computable from the information given in the first part. Hence, the initial rule of the grammar suitable for the generation of such problems is:

$$P \rightarrow G\,Q.$$

P is the initial non terminal symbol with the meaning "problem." This symbol is rewritten by the sequence $G\,Q$ of non terminal symbols, where G stands for the part of the problem with the given facts and Q represents the question based on G. This initial rule determines the fundamental frame for each problem that can be generated by means of the grammar. It is appropriate to represent each sentence[1] of the first part G of the problem by a separate non terminal symbol, because in each sentence a variable information is provided. Hence, A stands for sentence (1), B for sentence (2) and C for sentence (3). By the rewriting rule $Q \rightarrow D \mid E$ a new notation is introduced. This expression means that Q may either be rewritten by D or E. From this rule it can be seen that applying this grammar, a *set* of problems can be constructed, where each problem of this set is alternatively characterized by question-parts of type D or type E. Next, consider the rewriting rule for the non terminal symbol A:

$$A \rightarrow \text{[A grocer bought] } N_1 \text{ [dozen pears for \$] } N_2 \text{ [.]}.$$

By convention, terminal symbols like '[A grocer bought]' are written in parentheses. This rule shows that symbols can be rewritten by sequences that consist both of terminal and non terminal symbols. N_1 and N_2 are rewritten as follows:

$$\begin{aligned} N_1 &\rightarrow x_1(x_1 \in \boldsymbol{N} \wedge 5 \leq x_1 \leq 20), \\ N_2 &\rightarrow x_2(x_2 \in Q^+ \wedge 1 \leq x_2 \leq 8 \wedge x_2 \leq \tfrac{x_1}{1.5}). \end{aligned}$$

By these two rules, a further convention for the notation of rewriting is introduced. The non terminal symbols N_1 and N_2 are rewritten by numerical values

[1] The term 'sentence' describes in this case a sentence of the English language and *not* a sentence of the 'problem language' being just defined. A sentence in this language is per definition constituted by a complete problem.

that have to be within a range shown in parentheses. Formally, the rules that assign ranges of values to symbols are equivalent to rules that determine the rewriting by an "or"-expression (e. g., $Q \rightarrow D \mid E$). Defining ranges, however, helps to abbreviate the formulation of rules. Also, for simplifying the notation, in many cases terminal symbols consist of strings (e. g. [A grocer bought]). Because these strings represent problem elements that are either left unchanged or changed as a whole, it is not necessary to split them up into separate terminal symbols. The complete grammar and the vocabularies V_T and V_N for a problem set to which the example problem belongs are shown in Table 1.

The question now is, what is gained from the definition of a problem grammar. This is subject matter of the following section, where a way of problem coding is introduced that is based on the existence of a grammar. In short, the interrelationship between grammars and problem codings can be characterized as follows:

1. Given a problem, it is desired to be able to assign a *unique* coding to the problem.[2] This can be achieved by determining relationships between elements of the grammar and elements of the code.

2. Given a problem coding, it should be possible to generate one or more problems (i. e., problem texts) that *uniquely* correspond to the coding. This can be important, for example, for automatic diagnostic procedures that work on the basis of problem codings. On the basis of grammars, problem generation can principally be conducted by computers. The existence of a well-developed grammar and the interrelationship between coding and grammar can serve to facilitate automatic problem generation during a computerized diagnostic process.

PROBLEM CODING

As already mentioned in the previous section, it is assumed that the difficulty of problems can be varied through modifications of special parts of problems. By determining such modifications an important prerequisite for deriving knowledge structures is established. In general constructing a problem grammar is seen as basic for problem coding and imposing structures on problem sets. Hence, with an inappropriate grammar it may be impossible to establish an appropriate coding or problem structure.

The first question now is, what elements of problems shall be coded. We can, on the one hand, take *textual or symbolic sequences* appearing within the problem formulation. According to this point of view, one may split up a problem into units that correspond to the most characteristic parts of the formulation of a problem. On the other hand the *knowledge or operations* necessary for solving a problem correctly might be regarded as relevant problem elements.

[2] Note that the codings will be necessary for deriving problem structures.

TABLE 1
The Complete Grammar for a Problem Set

		Vocabularies
V_N	=	{**P**, G, Q, A, B, C, D, E, F, H, N_1, N_2, N_3, N_4, N_5}
V_T	=	{ [A grocer bought], [dozen pears for $], [dozen of the pears spoiled.], [The grocer wants to make a profit equal to], [of the total cost.], [At what price], [must he sell the remaining pears?], [By what amount has the price (per dozen) to be increased because of the spoiled pears?], [percent], [per dozen], [per pear], [.], $x_1(x_1 \in \boldsymbol{N} \wedge 2 \leq x_1 \leq 20)$, $x_2(x_2 \in Q^+ \wedge 1 \leq x_2 \leq \frac{x_1}{1.5})$, $x_3(x_3 \in \boldsymbol{N} \wedge 1 \leq x_3 < x_1)$, $x_4(\frac{1}{10} \leq x_4 \leq \frac{8}{10} \wedge x_4$ is a regular fraction), $x_5(x_5 \in \boldsymbol{N} \wedge 10 \leq x_5 \leq 80)$}
		Rules
P	→	G Q
G	→	A B C
Q	→	D | E
A	→	[A grocer bought] N_1 [dozen of pears for $] N_2 [.]
B	→	N_3 [dozen of the pears spoiled.]
C	→	[The grocer wants to make a profit equal to] F [of the total cost.]
D	→	[At what price] H [must he sell the remaining pears?]
E	→	[By what amount has the price (per dozen) to be increased because of the spoiled pears?]
F	→	N_4 | N_5 [percent]
H	→	[per dozen] | [per pear]
N_1	→	$x_1(x_1 \in \boldsymbol{N} \wedge 2 \leq x_1 \leq 20)$
N_2	→	$x_2(x_2 \in Q^+ \wedge 1 \leq x_2 \leq \frac{x_1}{1.5})$
N_3	→	$x_3(x_3 \in \boldsymbol{N} \wedge 1 \leq x_3 < x_1)$
N_4	→	$x_4(\frac{1}{10} \leq x_4 \leq \frac{8}{10} \wedge x_4$ is a regular fraction)
N_5	→	$x_5(x_5 \in \boldsymbol{N} \wedge 10 \leq x_5 \leq 80)$

Both of those points of view have been supported in numerous theoretical approaches. Textual or symbolic sequences as problem elements occur, for instance, in the *Facet Theory* of Guttman (1957), or in the investigations of Herbst (1978). Elements of knowledge or operations are considered as characteristic parts of problems in the *Structural Learning Theory* of Scandura (1973) and a model of Scheiblechner (1972), to name just a few.

The approach introduced here combines those points of view. Textual or symbolic elements of problems are associated with elements of knowledge and operations necessary for solving the problems. In this respect, the presented principle for problem coding is both a further development of, and an addition to, the ideas of Albert and Held (1994). The grammar ensures that each of the problems in a generated "problem universe" is correct in a syntactic and semantic sense. In order to distinguish between problems and problem codings that are introduced below, the following is defined:

DEFINITION 4.3 A *set of problems* is completely described by a problem language $L_P(G_P)$ with G_P the problem grammar.[3] For further information see Held (1993). The grammar G_P consists of a quadruple $G_P = (V_{TP}, V_{NP}, R_P, P)$ with V_{TP} the terminal vocabulary, V_{NP} the non terminal vocabulary, R_P a set of rules and P the initial non terminal symbol with the meaning of "problem". □

Hence, by the term problem, it is referred to a sequence of terminal strings that are rewritable for the non terminal symbol P of a problem grammar. In order to establish a *problem coding*, which provides different codes for problems of different difficulty, the concepts of *problem components* and *problem attributes* are introduced. Components and attributes are the *codes* that label those non terminal and terminal symbols of a problem grammar that are assumed to be related to the difficulty of the problems.

DEFINITION 4.4 (*Problem coding*) Let $G_P = (V_{NP}, V_{TP}, R_P, P)$ a problem grammar and $C = \{a, \ldots, n\} \cup \{\ell\}$ a set of "coding symbols." Symbol ℓ is an "empty coding symbol." The *coding of a problem component* is established by a mapping $\kappa : 2^{V_{NP}} \setminus \{\emptyset\} \to C$. By κ it is determined whether a subset of V_{NP} is coded by an element of C that is different from the empty coding symbol ℓ, or not.[4] A subset of V_{NP} coded by an element i of $C \setminus \{\ell\}$ is denoted v_i.

Furthermore, let for each coding symbol i in $C\setminus\{\ell\}$ exist a set $C_i = \{i_1, i_2, \ldots, i_j\} \cup \{\ell_0\}$ of coding symbols together with the empty coding symbol ℓ_0 and let the set of terminal symbols rewritable for the non terminal symbols in v_i (with respect to grammar G_P) be denoted V_{Ti}.

[3] Note that L_P is a so-called "type-1 language."

[4] The coding by an empty coding symbol is introduced for technical reasons: Each subset of a vocabulary coded by an empty symbol is considered as not relevant for representing a problem by a problem code.

The *coding of a problem attribute* is established by a mapping $\alpha : 2^{V_{Ti}} \setminus \{\emptyset\} \to C_i$. Hence, mapping α determines the coding of elements of $2^{V_{Ti}}$. A subset of V_{Ti} coded by an element i_1 of $C_i \setminus \{\ell_0\}$ is denoted v_{i_1}.

The set of *problem components* corresponding to a problem set is defined by mapping κ. The set of *problem attributes* of a component i is defined by mapping α. A component i is the set of attributes $i_1, \ldots, i_n$ corresponding to i with respect to mapping α. □

The relationship between components and attributes is determined by the rules in R_P. Given a problem grammar, each element of the power set of V_{NP} is a potential component. The selection of the appropriate subsets of V_{NP} by which components are defined is dependent on the problem type and on the intended problem variations. The following example illustrates the principles of problem coding as introduced in Definitions 4.3 and 4.4.

EXAMPLE 4.1 Consider again the "grocer problem" on page 70. The grammar shown in Table 1 (p. 72) provides the complete description of the problem language and hence of a problem set. As defined above, components are inferred from subsets of the non terminal vocabulary. For the example, the non terminal symbols Q, F, N_1, N_2, N_3 are selected for defining components. As shown in the following, those symbols are assumed to be involved with variations of the difficulty of the generated problems. Symbol Q represents the "question part" of the problem. This non terminal symbol can be rewritten by three different strings of terminal symbols:

[At what price][per dozen][must he sell the remaining pears?],
[At what price][per pear][must he sell the remaining pears?],
[By what amount has the price (per dozen) to be increased because of the spoiled pears?].

Symbol F is "embedded" in the string by which symbol C is rewritten. Corresponding to the grammar two types of terminal strings can be inserted for F:

x_4,
x_5 [percent].

The possible values of x_4 and x_5 are determined by the rewriting rules for N_4 and N_5. Rewriting F either by x_4 or by x_5 [percent] determines whether the problem solver is confronted with a regular fraction or a percentage. The terminals belonging to N_1, N_2 and N_3 can be selected from the respective sets given in the grammar.

For the example problems, three components are defined that are associated with the non terminals outlined above. Component a stands for the question sentence of the problems and is therefore based on symbol Q. Each of the three sentences rewritable for Q corresponds to one of the attributes a_1, a_2 and a_3 of

a. The second component b represents the alternative formulations concerning percentages and regular fractions. The corresponding symbol is F. Calculating a percentage is coded by attribute b_1, whereas the regular fractions are coded by b_2. Finally, component c covers the range of numbers relevant for calculating the result. This range is fixed by symbols N_1, N_2, and N_3. Three different ranges are denoted by attributes c_1, c_2, and c_3. □

Problems as a whole are coded as n-tuples of attributes. These n-tuples can be generated by forming the *Cartesian product* of components. Consider the continuation of Example 4.1.

EXAMPLE 4.2 In the previous example, components $a = \{a_1, a_2, a_3\}$, $b = \{b_1, b_2\}$, and $c = \{c_1, c_2, c_3\}$ have been established. The Cartesian product $a \times b \times c$ yields a set $\mathcal{P}$ of 18 3-tuples of attributes:

$$\begin{aligned}\mathcal{P} = \{&(a_1,b_1,c_1),(a_2,b_1,c_1),(a_3,b_1,c_1),(a_1,b_2,c_1),(a_2,b_2,c_1),(a_3,b_2,c_1),\\ &(a_1,b_1,c_2),(a_2,b_1,c_2),(a_3,b_1,c_2),(a_1,b_2,c_2),(a_2,b_2,c_2),(a_3,b_2,c_2),\\ &(a_1,b_1,c_3),(a_2,b_1,c_3),(a_3,b_1,c_3),(a_1,b_2,c_3),(a_2,b_2,c_3),(a_3,b_2,c_3)\}.\end{aligned}$$

Consider two problems coded (a_1, b_1, c_1) and (a_3, b_2, c_2):[5]

(a_1, b_1, c_1):

(1) A grocer bought $\overset{c}{\boxed{3}}$ dozen of pears for \$ $\overset{c}{\boxed{1.99}}$.

(2) $\overset{c}{\boxed{1}}$ dozen of the pears spoiled.

(3) The grocer wants to make a profit equal to $\overset{b}{\boxed{\tfrac{1}{2}}}$ of the total cost.

(4) $\overset{a}{\boxed{\text{At what price per dozen must he sell the remaining pears?}}}$

(a_3, b_2, c_2):

(1) A grocer bought $\overset{c}{\boxed{19}}$ dozen of pears for \$ $\overset{c}{\boxed{12.49}}$.

(2) $\overset{c}{\boxed{8}}$ dozen of the pears spoiled.

(3) The grocer wants to make a profit equal to $\overset{b}{\boxed{\text{35 percent}}}$ of the total cost.

(4) $\overset{a}{\boxed{\text{At what price per pear must he sell the remaining pears?}}}$

□

[5] The strings corresponding to components are marked by framed boxes.

PROBLEM STRUCTURES

Having defined the concept of a problem, we now show how surmise relations and surmise systems can be imposed on component-based problem sets.[6] At first, ordering principles that directly operate on problem components and attributes are introduced. The main prerequisite for application are structural properties of components. Further, we demonstrate how these methods are related to "demands" or operations that are supposed to be prerequisites for solving the problems.

Surmise relations

The central principle described here is the "*componentwise product order.*" It is shown how this order can be derived from problem components, and how the relation can be "weakened" by systematically removing ordered pairs. For establishing such orders each possible pair of problems in a given set has to be compared "component by component." For such a comparison, assumptions about a relation on the attributes of a component are required. These assumptions can formally be expressed as orders on the respective attribute sets.

DEFINITION 4.5 A partial order which is defined on a set $a = \{a_1, \ldots, a_n\}$ of attributes of a problem component a is called *attribute order* $\langle a, \trianglelefteq_a \rangle$. □

Within the further considerations, attribute orders are supposed to be transitive, reflexive, and antisymmetric orders (i. e., partial orders). Generally, the assumed order depends on the focused knowledge domain and on the selected components and attributes. The foundation of attribute orders from a psychological point of view is discussed later.

The order on a problem set is established by a "componentwise"' comparison of attributes. This is equivalent to ordering products of sets, which is known from ordering theory (Birkhoff, 1973).

DEFINITION 4.6 Let $M_1, \ldots, M_n$ be sets on that the reflexive, transitive, and antisymmetric relations $\trianglelefteq_1, \ldots, \trianglelefteq_n$ are defined. On the Cartesian product $M_1 \times \cdots \times M_n$ a *componentwise relation* $\preceq$ is imposed by the following condition:

$$(x_1, \ldots, x_n) \preceq (y_1, \ldots, y_n) \Leftrightarrow (\forall i) x_i \trianglelefteq_i y_i \text{ in } M_i.$$

□

Note that this definition holds for partially ordered sets $M_1, \ldots, M_n$. The resulting order on the product is also a partial order (Birkhoff, 1973). With respect to the problem grammar by which the components and attributes have been induced,

[6] Methods for establishing component-based surmise relations have first been outlined by Albert (1989).

possibly some of the attribute tuples do not correspond to constructible problems. Therefore, it might be necessary to remove all ordered pairs from $\preceq$ which contain such an attribute tuple. The componentwise relation on problem sets introduced in Definition 4.6 can be paraphrased as follows: If a subject solves problem A, it is surmised that he or she is also capable of solving problem B, if and only if all attributes in A are at least as "difficult" as the corresponding attributes in B with respect to the given attribute orders.

It appears to be plausible that a correct solution to a problem implies a correct solution to this very problem. This case is covered by the reflexivity of $\preceq$. How "different" two problems that solely differ with respect to one attribute really are, depends on the kind of selected attributes. If we consider the "grocer problems" shown in the previous sections, it is easy to imagine that in some cases the difference between two differently coded problems is not as pertinent as it might be desired for considering this pair as an element of a valid surmise relation. Hence, it may be reasonable to postulate that ordered pairs of problems either have to consist of equivalent problems, or have to contain problems that are different with respect to at least m attributes. Aside from the number of different ("super-ordinate") attributes, one can also focus on the number of attributes that are "situated between" the compared attributes. We define:

DEFINITION 4.7 Given a component $a = \{a_1, a_2, \ldots, a_n\}$, for a pair $(a_i, a_j) \in a \times a$ of attributes, the *span* s is obtained by determining the number of attributes a_k for that $a_i \trianglelefteq_a a_k$ and $a_k \trianglelefteq_a a_j$ (with $a_i, a_j \neq a_k$) hold. If there exists more than one linear order in $\trianglelefteq_a$ with endpoints a_i and a_j, the *minimal* linear order is selected for determining s. □

Based on the numbers m and s, a relation $\preceq_{\delta_n}$ can be obtained by removing all ordered pairs from $\preceq$ (see Definition 4.6) that do not stand out due to a condition that is stated by the following definitions.

DEFINITION 4.8 Let the ordered pair (X, Y) be an element of a componentwise relation $\preceq$. The number of components in X and Y that are equipped with different attributes are called *componentwise distance* $m_{X,Y}$ of (X, Y). Let m_j be the distance between the attributes of component j in problem X and Y; $m_j = 0$ if the attributes of component j are identical, $m_j = 1$ if the attributes differ. Then the componentwise distance $m_{X,Y}$ for problems X and Y associated with n components is determined by $m_{X,Y} = \sum_{j=1}^{n} m_j$. Furthermore, the *componentwise span* of (X, Y) is obtained by adding the spans for all components in X and Y. Let s_j be the span between attributes in component j. Then the componentwise span $s_{X,Y}$ for problems X and Y associated with n components is determined by $s_{X,Y} = \sum_{j=1}^{n} s_j$. □

DEFINITION 4.9 Let $M_1, \ldots, M_i$ be partially ordered sets of attributes with attribute relations $\trianglelefteq_1, \ldots, \trianglelefteq_i$ defined on them. If on the Cartesian product

$M_1 \times \cdots \times M_i$ a componentwise order $\preceq$ (see Definition 4.6) is defined, a δ_n-*weakened* relation $\preceq_{\delta_n}$ is imposed on the product by removing all ordered pairs (X, Y) from $\preceq$ that do not fulfill one of the following conditions:

(Cond1) $m_{X,Y} + s_{X,Y} \geq n$,

(Cond2) $m_{X,Y} = s_{X,Y} = 0$ (i. e., X and Y are equivalent). □

All "reflexive" pairs of $\preceq$ are members of $\preceq_{\delta_n}$ and each element of $\preceq_{\delta_n}$ is also a member of $\preceq$. The relation $\preceq_{\delta_n}$ also imposes a partial order on $M_1 \times \cdots \times M_i$.

Generally, increasing n decreases the "strictness"[7] of the corresponding surmise relation. In terms of the corresponding knowledge structures, this means that the number of knowledge states grows with increasing n. In principle, each relation $\preceq_{\delta_n}$ is compatible with the assumptions formulated through the attribute orders. Which of the $\preceq_{\delta_n}$ relations is the most appropriate for a given body of problems is dependent on the focussed domain, on the selected components and attributes, and on the diagnostic process for which the problems are used. Clearly, weakening the structure yields less information about the influence of the specific demands on the difficulty of the problems. Also, with increasing n, assessment procedures will become more and more inefficient because the corresponding relations lead to increasing numbers of knowledge states. Conversely, a relation that is too strict may appear to be not valid, and hence not be appropriate for diagnostic use.

Surmise systems

The approach introduced in the previous section is designed for establishing surmise relations on sets of component-based problems. Surmise relations may be fully adequate for domains that are structured quite simply and free from more or less ambiguous relationships between the inherent "elements of knowledge". If, however, problems of a given set can be solved in alternative ways that are assumed to be related to different elements of knowledge, the establishment of a surmise relation may not be appropriate for structuring a body of knowledge. Hence, the case of component-based problems with alternative prerequisites has to be taken into account.

Our aim is to obtain a surmise system for a component-based set of problems through the prerequisite relationships defined on the attributes of the problems. Analogously to surmise mappings, for every attribute a_i in an attribute set the (minimal) set(s) of prerequisite attributes is (are) provided. We define:

DEFINITION 4.10 An *attribute mapping* is a function $\varphi_a : a \to 2^{2^a}$ that associates to each element a_i of a component a a family of subsets of a including a_i. These subsets are called *attribute clauses* for a_i. □

[7] The term *strictness of a relation* is a label for the relative number of knowledge states that correspond to this relation: The stricter the relation is, the less states correspond to it.

For purposes that are described below, it is postulated that for every set of attributes, the attribute clauses fulfill the three axioms for (space-like) surmise systems stated in Doignon and Falmagne (1985, p. 186). In terms of attributes, this means that (1) each attribute clause for an attribute a_i contains a_i, (2) if an attribute a_j is an element of some attribute clause K for a_i, then there must be some attribute clause K' for a_j included in K and (3) any two attribute clauses for a_i are pairwise incomparable with respect to inclusion.

Given the attribute mapping, the next step is the construction of a *surmise mapping* based on attributes. In general, a surmise mapping σ determines for each problem A in a problem set Q the minimal subsets of problems in Q which are supposed to be subordinate to A. In terms of Doignon and Falmagne (1985), these subsets are called *clauses*. For disassociation from attribute clauses, they will be called *problem clauses* in the following. The problem clauses associated with an attribute tuple $(a_i, b_j, \ldots, x_k)$ are obtained by forming the Cartesian products of all possible combinations of attribute clauses of $a_i, b_j, \ldots, x_k$: Consider a problem associated with an attribute tuple $(a_i, b_j, \ldots, x_k)$ that contains attributes out of components $a, \ldots, x$. For each of these attributes an attribute mapping is defined (i. e., $\varphi_a(a_i), \varphi_b(b_j), \ldots, \varphi_x(x_k)$). The attribute clauses of an attribute a_i are denoted K_{a_i}, K'_{a_i}, K''_{a_i}, and so on Hence, $\varphi_a(a_i) = \{K_{a_i}, K'_{a_i}, \ldots\}$. The product $\varphi_a(a_i) \times \varphi_b(b_j) \times \cdots \times \varphi_x(x_k)$ determines all possible tuples $(K_{a_i}, K_{b_j}, \ldots, K_{x_k})$, $(K'_{a_i}, K_{b_j}, \ldots, K_{x_k})$, ... of attribute clauses with respect to problem $(a_i, b_j, \ldots, x_k)$. From each of these tuples of attribute clauses, one clause for the problem is inferred by forming the product of all attribute clauses out of the tuple.

By this procedure, the surmise mapping $\sigma(A)$ for a problem A is determined. Consider an example:

EXAMPLE 4.3 Let $\mathcal{P}$ be a component-based problem set:

$$\begin{aligned}\mathcal{P} = \; & \{(a_1, b_1), (a_1, b_2), (a_1, b_3), (a_2, b_1), (a_2, b_2), (a_2, b_3), \\ & (a_3, b_1), (a_3, b_2), (a_3, b_3)\}\end{aligned}$$

The corresponding components are $a = \{a_1, a_2, a_3\}$ and $b = \{b_1, b_2, b_3\}$. Assume that the following attribute mappings are defined:

$$\begin{aligned}\varphi_a(a_1) &= \{\{a_1\}\}, & \varphi_b(b_1) &= \{\{b_1\}\}, \\ \varphi_a(a_2) &= \{\{a_2\}\}, & \varphi_b(b_2) &= \{\{b_1, b_2\}\}, \\ \varphi_a(a_3) &= \{\{a_1, a_3\}, \{a_2, a_3\}\}, & \varphi_b(b_3) &= \{\{b_1, b_2, b_3\}\}.\end{aligned}$$

The AND/OR Graph corresponding to φ_a and the Hasse diagram corresponding to φ_b look as shown in Fig. 1.

It can be seen that attributes a_1, a_2, b_1, b_2, and b_3 each have one attribute clause. Only attribute a_3 has two attribute clauses. Due to the procedure described above, the problem clauses for each problem in $\mathcal{P}$ can be determined. Consider,

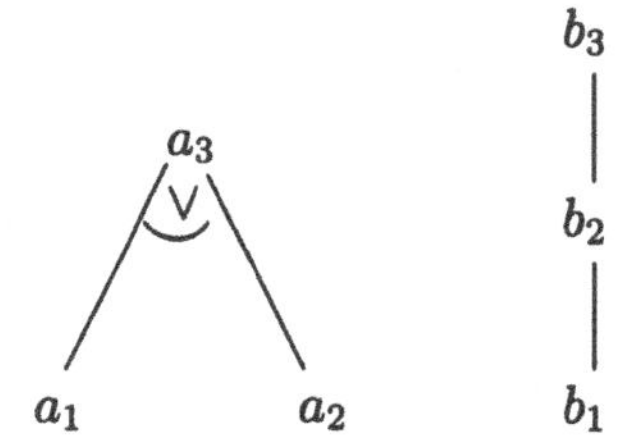

FIG. 1. The AND/OR Graph corresponding to φ_a and the Hasse diagram corresponding to φ_b.

for example, attribute tuple (a_3, b_1). Attribute mappings φ_a and φ_b determine the attribute clauses for attributes a_3 and b_1: $\varphi_a(a_3) = \{\{a_1, a_3\}, \{a_2, a_3\}\}$, $\varphi_b(b_1) = \{\{b_1\}\}$. We set $K_{a_3} = \{a_1, a_3\}$, $K'_{a_3} = \{a_2, a_3\}$, and $K_{b_1} = \{b_1\}$. The product $\varphi_a(a_3) \times \varphi_b(b_1)$ yields tuples (K_{a_3}, K_{b_1}) and (K'_{a_3}, K_{b_1}). The first clause for the problem associated with (a_3, b_1) is obtained by

$$K_{a_3} \times K_{b_1} = \{a_1, a_3\} \times \{b_1\} = \{(a_1, b_1), (a_3, b_1)\}.$$

The second and last clause is obtained by

$$K'_{a_3} \times K_{b_1} = \{a_2, a_3\} \times \{b_1\} = \{(a_2, b_1), (a_3, b_1)\}.$$

The problem clauses for the other problems in $\mathcal{P}$ are constructed in the same way.

□

Analogously to weakened surmise relations introduced above, weakening of component-based surmise functions can be performed. For this purpose, the problem clauses are reduced by those elements that do not fulfill a "criterion of difference" with respect to the considered problem. As with the $\preceq_{\delta_n}$ relation, this criterion is embodied by a numerical value that is the sum of two distance indices. Again, the *componentwise distance* (see Definition 4.8) is considered. Because the *componentwise span* introduced in Definition 4.8 is dependent on an attribute order, a new type of span that can be determined for pairs of problems in a component-based surmise system has to be defined.

DEFINITION 4.11 Given a component $a = \{a_1, \ldots, a_n\}$ and an attribute mapping φ_a by which for each attribute in a the set of prerequisites is determined. For each pair $(a_i, a_j) \in a \times a$ with $a_i \neq a_j$, the *span* s is obtained by counting the attributes a_k for that $a_k \in \varphi_a(a_i)$ and $a_j \in \varphi_a(a_k)$, with $a_i, a_j \neq a_k$, holds.

□

Based on this, the componentwise distance and the componentwise span for problems in a component based surmise system are defined.

DEFINITION 4.12 Let problem X be an element of a clause of problem Y (i. e., $\sigma(Y) = \{\{\ldots, X, \ldots\}, \ldots\}$) with $X \neq Y$. The *componentwise distance* between X and Y is given by the number of components associated with X and Y that are equipped with different attributes. Let m_j be the distance between the attributes of component j associated with problems X and Y; $m_j = 0$ if the attributes of component j are identical, $m_j = 1$ if the attributes differ. Then the componentwise distance $m_{X|Y}$ for problems X and Y associated with n components is determined by $m_{X|Y} = \sum_{j=1}^{n} m_j$. The *componentwise span* between X and Y is obtained by adding the spans for all corresponding components in X and Y. Let s_j be the span between attributes in component j in problems X and Y. Then the componentwise span $s_{X|Y}$ for problems X and Y associated with n components is determined by $s_{X|Y} = \sum_{j=1}^{n} s_j$. □

The "weakened" surmise systems are obtained as follows:

DEFINITION 4.13 Let Q be a problem set and σ a surmise function defined on the problems in Q. A δ_n*-weakened* surmise function σ_{δ_n} is obtained by removing each problem from the clauses of the problems in Q that does not fulfill the following conditions. For $X, Y \in Q$ and X, an element of a clause in $\sigma(Y)$ it holds that X is also an element of a clause in $\sigma_{\delta_n}(Y)$ if and only if either

(Cond1) $m_{X|Y} + s_{X|Y} \geq n$, or

(Cond2) $m_{X|Y} = s_{X|Y} = 0$ (i. e., X and Y are equivalent). □

Attributes and problem demands

Up to this point, the question how the structural properties of attribute sets are obtained and founded has not been discussed. In the previous sections, those properties were paraphrased by the term "difference in difficulty." In many cases, it may—at least for experts—be obvious and plausible, which attribute of a component makes a problem as a whole more difficult than another attribute. Also, the question for the prerequisites that are needed for tackling with *demands* being evoked by some attributes may be answerable for experts. This section deals with the demands that are supposed to be associated with attributes, and it is shown how demands are used for the establishment of attribute orders and attribute mappings.

In Falmagne, Koppen, Villano, Doignon, and Johannesen (1990) a concept for assigning "skills" to problems is sketched. The ideas introduced there serve as background for the following considerations. It is assumed that a set S of skills is given, and that for each element of a problem set, there exists a subset of S with the skills required for solving the problem. A binary relation can be imposed on the problem set by inclusion of the corresponding skill sets. This relation is an equivalent to the surmise relation. If multiple sets of skills (the "competencies") are assigned to a problem, then families of knowledge states can be generated by collecting all states that are "compatible" with a competency (cf. Falmagne et al., 1990; Definition 3).

In contrast to this approach, skills or demands are in our approach not assigned to problems as a whole—the attention is directed to demands that are associated with the attributes of a component. This is convenient with our general view of problems: the definition of components is framed in order to be able to control the variable problem factors which might be relevant for the solution process. The assumptions regarding the relationship between problem variation, and the solution process, are expressed by assigning demands to attributes. If a component is determined, the critical question is, which attributes have to be defined in order to ensure that they are associated with different sets of demands.

By the following definitions it is determined how *attribute orders* can be derived from assumptions regarding demands. First of all, "demand assignments" are defined. Suppose that for each component a, a set $\mathcal{D}_a$ of "component-specific" demands exists. Note that the following definition, as well as most of the following formal statements, is analogous to the considerations concerning "skill assignments" in Doignon (1994).

DEFINITION 4.14 Let a be a component, and $\mathcal{D}_a$ a set of demands. A mapping $\rho_a : a \to 2^{\mathcal{D}_a} \setminus \{\emptyset\}$ determines a *demand assignment* for a. This means that for each attribute $a_n \in a$, there exists a nonempty subset of $\mathcal{D}_a$, that is assigned to a_n. A demand assignment is denoted $(a, \mathcal{D}_a; \rho_a)$. □

An attribute order is established as follows:

DEFINITION 4.15 An *attribute order* $\trianglelefteq_a$ is *imposed* on a component a by the condition

$$a_i \trianglelefteq_a a_j \Leftrightarrow \rho_a(a_i) \subseteq \rho_a(a_j), \text{ with } a_i, a_j \in a.$$

□

According to Definition 4.15, attribute orders are based on inclusion of demand sets. The condition formulated above means that a pair (a_i, a_j) of attributes is an element of an attribute relation $\trianglelefteq_a$ if and only if the set of demands assigned to attribute a_i is a subset of the set of demands assigned to attribute a_j. As is known from set theory, the relation $\subseteq$ of set inclusion imposes a partial order (reflexive, transitive, antisymmetric) on a family of sets. Based on such orders, a surmise relation on a problem set can be derived.

As a next step, we show how attribute mappings can be derived by demand assignments. An attribute mapping determines a set of prerequisite sets—the attribute clauses—for each attribute in a component. The basic idea for demand assignments is that multiple sets of demands can be attached to an attribute. These sets are called *requirements* of an attribute. Multiple requirements can correspond to for example alternative possibilities for the solution of a problem or to alternative ways of processing the problem text. It is supposed that each requirement of an attribute contains all the demands that have to be fulfilled if this attribute is contained in a problem coding. A *demand multiassignment* is defined as follows:

DEFINITION 4.16 Given a component a, a *demand multiassignment* [8] is determined by a set $\mathcal{D}_a$ of demands corresponding to a, and a mapping $\omega_a : a \to 2^{2^{\mathcal{D}_a}\setminus\{\emptyset\}}\setminus\{\emptyset\}$ (i. e., a mapping from a into the set of all nonempty families of nonempty subsets of $\mathcal{D}_a$). A demand multiassignment is denoted as $(a, \mathcal{D}_a; \omega_a)$.

□

DEFINITION 4.17 The collection of *attribute clauses* for an attribute a_n "generated" by $(a, \mathcal{D}_a; \omega_a)$ is determined by collecting all subsets C of a that satisfy, for some subset S of $\omega_a(a_n)$:

$$C = \{a_i \in a \mid \exists \mathcal{R} \in \omega_a(a_i) \colon \mathcal{R} \subseteq S\}.$$

□

Expressed in words: a clause for attribute a_n consists of all $a_i \in a$ for that holds that there exists a requirement $\mathcal{R}$ in $\omega_a(a_i)$ that is a subset of one of the requirements in $\omega_a(a_n)$.

If the clauses for problems do not meet the three conditions for (space-like) surmise systems, then the corresponding knowledge states may neither be $\cup$-closed, nor $\cap$-closed. This is not really problematical if the properties of the knowledge spaces are not required explicitly, as this is, for example, the case for the adaptive diagnostic procedures introduced by Falmagne and Doignon (1988a, 1988b).

EXPERIMENTAL INVESTIGATION

In this investigation we make use of theoretical concepts developed in the previous sections. It is shown how a set of problems in the domain of "elementary stochastics" is constructed, and how problem structures are imposed on this set.

First, the focused *domain*, and the *type of problems* used in the investigation are described. Then, the *problem grammar* is established and, based on the grammar, the *problem components* and their *attributes* are defined. The *demand assignment* we propose is both based on curricular or "logical" properties inherent to the domain and on properties of the wording of the problems. Aside to a *main model*, also *alternative models* are tested, by which the influence of single components on solution behavior can be investigated.

[8] This definition is analogous to the definition of "skill multiassignments" in Doignon (1994). Based on these assignments, knowledge structures can be "delineated." Note that the definition of demand multiassignments given here uses some of Doignon's terminology within a different context, actually for determining the generation of attribute clauses.

Knowledge domain and type of problems

The knowledge domain to be investigated can vaguely be termed "elementary stochastics." Generally, stochastics covers the areas of probability theory, and mathematical statistics. The knowledge to be assessed in the investigation is a very small "subset" of this domain; in fact, only the "first few pages" of usual stochastics textbooks are relevant. Exclusively simple random experiments with finite sets of possible (convenient) events are considered. Because all elementary events are supposed to be equally probable, knowledge about the "classical probability space" due to Laplace is appropriate and also sufficient. In addition, some combinatorial principles, like the computation of the number of n-element subsets of a given set, have to be known. In most cases, however, no combinatorial formulas are required explicitly because the number of convenient events can also be obtained by "counting." The following problem can be regarded as prototypical:

(1) An urn contains three red and three blue balls.
(2) Two balls are drawn successively.
(3) Drawing is performed with replacement.
(4) The drawn balls are red.
(5) Compute the probability of this event.

Derivation of problem demands

The demands on which structural assumptions concerning the attributes (i. e., attribute orders and attribute mappings) are based are of two different types. On the one hand, there are demands concerning the mathematical content of the problems (e. g. mathematical propositions, rules for calculations). On the other hand, *comprehending the problem texts* is seen as meaningful demand on the problem solver. As we outline below, the problems generated for this investigation differ with respect to properties that can both be characterized in syntactic and semantic terms. All in all, three types of wordings are used that correspond to different grades of comprehensibility. First, however, consider Table 2, which shows the elements of the set $\mathcal{D}$ of demands relevant for the investigation. Although the mathematical demands require no further explanations, the assumptions that lead to a distinction between texts of different comprehensibility have to be discussed in greater detail. Two approaches that deal with variable aspects of problems in rather different (theoretical) respects are made use of. Those aspects are labeled "textual coherence", and "textual cohesion." The approaches being focused on are a processing model for "text comprehension and text production" proposed by Kintsch and van Dijk (1978), and a graph theoretical approach for the representation of cohesion of texts raised by Brainerd (1977).

TABLE 2
List of Demands of Problems on Elementary Probability Calculus (Simple Urn-Experiment)

$\mathcal{O}_1$:	Generally, Laplace probabilities are computed by the quotient $\frac{\|A\|}{\|\Omega\|}$. A is the set of "convenient events" with respect to a random experiment, and Ω is the set of "possible events."
$\mathcal{O}_2$:	Determining $\|\Omega\|$ for an elementary event (i. e. determining the total number of balls in the urn).
$\mathcal{O}_3$:	Determining the number of convenient events if only one ball is drawn.
$\mathcal{O}_4$:	Determining a convenient event if only one ball is drawn, or if the sample for that the probability has to be computed, consists of equally colored balls.
$\mathcal{O}_5$:	If the probability for an outcome like "exactly/at least m of the drawn balls are of color x" is asked (with m smaller than the number of drawn balls), all possible sequences of drawing are convenient events.
$\mathcal{O}_6$:	$p(A \cup B) = p(A) + p(B)$, with $A \cap B = \emptyset$; A and B are events.
$\mathcal{O}_7$:	$p(A \cap B) = p(A) \cdot p(B)$ holds if two events A and B are (stochastically) independent.
$\mathcal{O}_8$:	If there are different numbers of differently colored balls in the urn, the probability of drawing a ball of a specific color is not equal to 0.5.
$\mathcal{O}_9$:	Drawing *without* replacement decreases the total number of balls in the urn and also the number of balls that have the same color as the drawn ball.
$\mathcal{O}_{10}$:	If the probability for drawing "at least" a number of certain balls must be calculated, it has to be recognized that the appropriate event also covers results which are not explicitly stated in the question.
$\mathcal{O}_{11}$:	Comprehension of a text with wording version 1.
$\mathcal{O}_{12}$:	Comprehension of a text with wording version 2.
$\mathcal{O}_{13}$:	Comprehension of a text with wording version 3.

The model of Kintsch and van Dijk works with the "propositional representation" of text bases; it states how comprehension of the *meaning* of texts may be organized in memory. Brainerd's approach is oriented towards relationships between parts of "concrete" texts. The core of his considerations is a *dependence relation* being defined on the set of sentences in a text.

From the processing model of Kintsch and van Dijk, hypotheses concerning text (re-)production can be derived. An important prerequisite for this is the transformation of texts into *propositional text bases*. The problem with the transformations is that they are not necessarily unique. Brainerd's approach requires no transformation of texts; there is, however, no direct way proposed how empirical hypotheses about text comprehension may be established.

The aim of the considerations is to define a component "textual constitution", both involving textual coherence and cohesion, and to derive structures on the attributes of this component by using aspects of both of the approaches. Note that the theoretical approaches cannot be presented in sufficient detail within this paper. For a more elaborate description see Held (1993) and the original publications of Kintsch and van Dijk and Brainerd.

Coherence of Propositional Text Bases

The main question of Kintsch and van Dijk (1978) is how texts may be stored in, and retrieved from memory. The description of storage and retrieval is achieved by the formulation of a *cyclical processing model* which provides the steps being necessary for understanding the *meaning* of texts. The model works on the level of *propositions*, that are defined as units supposed to represent the meaning of a text. Due to Kintsch (1974) word concepts are the entries of the lexicon of semantic memory whereas the term "semantic memory" is treated as synonymous with "personal knowledge" (i. e., knowledge that is not culturally shared). Propositions are n-tuples of word concepts in the form $(P, A_1, \ldots, A_n)$, where P is called *predicate* and the A_i are called *arguments*. By means of rules for the propositional representation of textual meaning (Turner & Greene, 1977), texts as a whole can be transformed into lists of propositions. Such lists are called *propositional text bases*. From the text bases, a hierarchical semantic structure of a text can be derived, that means, it is supposed that relationships between the propositions exist. Due to these relationships, a text can be seen as a "coherent whole" that consists of interconnected units (the propositions). According to Kintsch (1974) and Kintsch and van Dijk (1978), coherence of a text is based on two properties: *argument overlap and embedding*. Those properties can be used in order to establish a *coherence graph*. Kintsch (1974) provides a set of five rules that are the basic "receipt" for transforming propositional text bases into coherence graphs. For an illustration of these topics consider the following problem together with the corresponding propositional text base and the coherence graph shown in Figure 2. Since the investigation has been conducted with German speaking subjects, we have to give examples in German language.

(1) In einer Urne befinden sich fünf weiße und fünf schwarze Kugeln.
(2) Aus der Urne werden nacheinander zwei Kugeln gezogen.
(3) Das Ziehen erfolgt ohne Zurücklegen.
(4) Die gezogenen Kugeln sind schwarz.
(5) Bestimmen Sie die Wahrscheinlichkeit für dieses Ereignis.

Given a propositional text-base and a coherence graph, the cyclic processing model of Kintsch and van Dijk (1978) can be applied. According to this model it is supposed that the number of propositions that can be processed at one time is not arbitrarily high. In principle, the model assumes the following cyclic process: A number of propositions (one "chunk") is loaded into working memory (WM), then a coherence graph is built from those propositions. The graph is transferred to long-time-memory, whereas some of its propositions stay in WM. Then, the

S1	P1	(UND P2 P5)
	P2	(SICH-BEFINDEN-IN KUGEL1 URNE)
	P3	(ANZAHL-VON KUGEL1 FÜNF)
	P4	(MOD KUGEL1 WEISS)
	P5	(SICH-BEFINDEN-IN KUGEL2 URNE)
	P6	(ANZAHL-VON KUGEL2 FÜNF)
	P7	(MOD KUGEL2 SCHWARZ)
S2	P8	(ZIEHEN $ KUGEL URNE)
	P9	(MOD P8 NACHEINANDER)
	P10	(ANZAHL-VON KUGEL ZWEI)
S3	P11	(ZIEHEN $ OHNE-ZURÜCKLEGEN)
S4	P12	(MOD KUGEL GEZOGEN)
	P13	(MOD P12 SCHWARZ)
S5	P14	(BESTIMMEN-FÜR $ WAHRSCHEINLICHKEIT P12)
	P15	(MOD P14 ANWEISUNG)

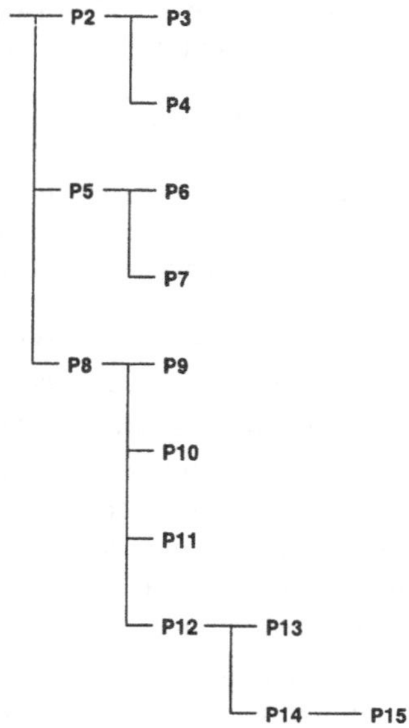

FIG. 2. Propositional text base and coherence graph of the example problem

next portion of propositions is loaded into WM and it is attempted to establish a coherence graph and to connect this graph to the remaining propositions of the previous graph. This procedure continues until all propositions have been loaded and a coherence graph of the complete propositional text-base has been built and transferred to long-time-memory. Difficulties arise when no connection between a new chunk and the propositions residing in WM can be made. Then time-consuming processes like *reinstatement searches* or *reorganizations* have to be performed.

Without going into further detail, we can state that for the experimental investigation three types of problem wordings have been selected that require—according to the model of Kintsch and van Dijk (1978)—different numbers and types of processes in order to establish a complete coherence graph.

Graph Theoretical Representation of Textual Cohesion

Alternatively to textual coherence, the textual cohesion of the problem texts also has been determined. It is important to note that in linguistic and psycholinguistic research, the terms "coherence" and "cohesion" are frequently used with intermixed meanings. In general, "coherence" mostly occurs in conjunction with relationships between abstract units of textual meaning, whereas "cohesion" is associated with syntactical and semantical relationships between textual units (i. e., the surface structure of texts). For example, Marcus (1980) states:

> "[...] the distinction between cohesion and coherence is described as follows: cohesion belongs to the text, i. e. to the semantic level of analysis: coherence belongs to the discourse, i. e. to the pragmatic level. Thus, cohesion is an immanent property of the text, whereas coherence refers to the text in a given situation." (p. 101)

For this investigation, the approach of Brainerd (1977) for the graph theoretical representation of textual cohesion has been used. Brainerd established a dependence relation between the sentences of a text. This relation is based on the five "Interpretations of semantic dependence" proposed by Lipski (1974). According to Brainerd, a text is *cohesive* if and only if it is *graph theoretically connected* with respect to some dependence relation. In principle, the degree of cohesion of a text is given by the number of edges that have to be deleted from a *cohesion graph* in order to make this graph unconnected. Note that in Held (1993), a more elaborate algorithm for determining the degree of cohesion is introduced. Finally, it is important to state that cohesiveness of problem texts is considered for supporting the assumptions resulting from the model of Kintsch and van Dijk (1978). Only if some text A is both less coherent and less cohesive than a text B, it is assumed that comprehending text A is more demanding than comprehending text B.

Problem grammar and coding

The grammar for the problems has to provide the possibility of generating different textual versions of problems. In general, the grammar admits variation of the following topics:

- Number of balls in the urn; ratio between the numbers of differently colored balls.
- Way of drawing: drawing once, drawing with or without replacement.
- Specification of the event: Each ball in the sample has the same color, exactly m of n (with $m < n$) drawn balls have a specific color, at least m of n (with $m < n$) drawn balls have a specific color.
- Textual variation: three kinds of wording.

The complete grammar is shown in Table 3. The problems being constructible with the grammar are supposed to be characterized by four components. Those components are defined via symbols of the non terminal vocabulary V_N.

The corresponding attributes are associated with terminal symbols in V_T rewritable for those non terminal symbols. In Table 4, an overview about all components and attributes is shown. The product $a \times b \times c \times d$ of problem components

TABLE 3
Problem Grammar for the Investigation

Vocabularies		
V_N	=	{**P**, G, Q, Q_1, Q_2, A, B, C, D, E, F, H, I, J, K, L, M, O, R, S, T, U, V, W, X, Y, N_1, N_2, N_3, N_4}
V_T	=	{[Bestimmen Sie die Wahrscheinlichkeit für dieses Ereignis], [Bestimmen Sie die Wahrscheinlichkeit], [In einer Urne befinden sich], [und], [Kugeln.], [Aus der Urne wird eine Kugel gezogen], [Die gezogene Kugel ist], [.], [Aus der Urne werden nacheinander], [Kugeln gezogen.], [Das Ziehen erfolgt], [Zurücklegen.], [Die gezogenen Kugeln sind], [der gezogenen Kugeln sind], [Eine wird gezogen.], [Die Kugel ist], [Es werden nacheinander], [gezogen.], [Es wird eine Kugel gezogen.], $t_1, t_2 \in$ {[weiß], [schwarz], [rot], [blau], [grün], [gelb]}, [e], [mit], [ohne], [Mindestens], [Genau], $x_1(x_1 \in N \wedge 1 \leq x_1 \leq 10)$, $x_2(x_2 \in N \wedge 1 \leq x_2 \leq 10)$, $x_3(x_3 \in N \wedge 2 \leq x_3 < (x_1 + x_2))$, $x_4(x_4 \in N \wedge 2 \leq x_4 < x_1, x_2)$}

Rules					
P	→	G Q	G	→	A | B | C
A	→	D E F | D H I J | D H I K	B	→	D L M | D O I J | D O I K
C	→	R M S | T I J S | T I K S	Q	→	Q_1 | Q_2
Q_1	→	[Bestimmen Sie die Wahrscheinlichkeit für dieses Ereignis]			
Q_2	→	[Bestimmen Sie die Wahrscheinlichkeit]			
D	→	[In einer Urne befinden sich] N_1 U [und] N_2 V [Kugeln.]			
E	→	[Aus der Urne wird eine Kugel gezogen]			
F	→	[Die gezogene Kugel ist] W [.]			
H	→	[Aus der Urne werden nacheinander] N_3 [Kugeln gezogen.]			
I	→	[Das Ziehen erfolgt] X [Zurücklegen.]			
J	→	[Die gezogenen Kugeln sind] W [.]			
K	→	Y N_4 [der gezogenen Kugeln sind] W [.]			
L	→	[Eine wird gezogen.]	M	→	[Die Kugel ist] W [.]
O	→	[Es werden nacheinander] N_3 [gezogen.]			
R	→	[Es wird eine Kugel gezogen.]			
S	→	[In der Urne befinden sich] N_1 U [und] N_2 V [Kugeln.]			
T	→	[Es werden nacheinander] N_3 [Kugeln gezogen.]			
U	→	$t_1 \in$ {[weiß], [schwarz], [rot], [blau], [grün], [gelb]}[e]			
V	→	$t_2 \in$ {{[weiß], [schwarz], [rot], [blau], [grün], [gelb]} \ t_1}[e]			
W	→	t_1 | t_2	X	→	[mit] | [ohne]
Y	→	[Mindestens] | [Genau]	N_1	→	$x_1(x_1 \in N \wedge 1 \leq x_1 \leq 10)$
N_2	→	$x_2(x_2 \in N \mid 1 \leq x_2 \leq 10)$			
N_3	→	$x_3(x_3 \in N \mid 2 \leq x_3 < (x_1 + x_2))$			
N_4	→	$x_4(x_4 \in N \mid 2 \leq x_4 < x_1, x_2)$			

TABLE 4
Overview About Components and Attributes of the Problems of the Investigation

Components	*Attributes*		
(a) numerical ratio of differently colored balls	(a_1) equal to one	(a_2) not equal to one	
(b) the way of drawing	(b_1) one-off	(b_2) with replacement	(b_3) without replacement
(c) specification of the asked event	(c_1) only equally colored balls are drawn	(c_2) exactly m of the drawn balls have the same color	(c_3) at least m of the drawn balls have the same color
(d) wording	(d_1) version 1	(d_2) version 2	(d_3) version 3

yields a total of 54 4-tuples of attributes $((a_1, b_1, c_1, d_1), \ldots, (a_2, b_3, c_3, d_3))$. Because 12 of these tuples can, with respect to the grammar, not be realized as problems, 42 tuples are relevant for the investigation.[9] Corresponding to each of these tuples, a number of (equivalent) problems can be derived from the grammar. For the investigation, 18 tuples out of 42 are selected. For each of them one problem has been constructed.

Hypothetical problem structures

At first a basic hypothesis is established, and then two alternative models. The main attention is directed to effects of textual variations (i. e., variations of textual cohesion and coherence) on solution performance. Concerning the textual component, a demand multiassignment is also proposed (Model 2). In all other models, however, the demand assignments lead to surmise relations, and therefore to quasi-ordinal knowledge spaces.

[9] The combination of attribute b_1 with attributes c_2 and c_3 is not possible because according to b_1 only one ball is drawn, whereas c_2 and c_3 refer to samples with more than one ball.

Model 1

It is supposed that the following demand assignments hold:

$$\begin{aligned} \rho_a(a_1) &= \{\mathcal{O}_1, \mathcal{O}_2\}, & \rho_b(b_1) &= \{\mathcal{O}_1, \mathcal{O}_3\}, \\ \rho_a(a_2) &= \{\mathcal{O}_1, \mathcal{O}_2, \mathcal{O}_8\}, & \rho_b(b_2) &= \{\mathcal{O}_1, \mathcal{O}_3, \mathcal{O}_7\}, \\ & & \rho_b(b_3) &= \{\mathcal{O}_1, \mathcal{O}_3, \mathcal{O}_7, \mathcal{O}_9\}, \end{aligned}$$

$$\begin{aligned} \rho_c(c_1) &= \{\mathcal{O}_1, \mathcal{O}_4\}, & \rho_d(d_1) &= \{\mathcal{O}_{11}\}, \\ \rho_c(c_2) &= \{\mathcal{O}_1, \mathcal{O}_4, \mathcal{O}_5, \mathcal{O}_6\}, & \rho_d(d_2) &= \{\mathcal{O}_{11}, \mathcal{O}_{12}\}, \\ \rho_c(c_3) &= \{\mathcal{O}_1, \mathcal{O}_4, \mathcal{O}_5, \mathcal{O}_6, \mathcal{O}_{10}\}, & \rho_d(d_3) &= \{\mathcal{O}_{11}, \mathcal{O}_{12}, \mathcal{O}_{13}\}. \end{aligned}$$

To each of the attributes, a single set of demands is assigned. Hence, *attribute orders* being founded by inclusion of the corresponding sets of demands can be formulated. The componentwise relation $\preceq_{\delta_1}$ (i. e. the δ_1-weakened surmise relation) imposed on the problem set is shown in Figure 3. With respect to the 18 problems, a total of $2^{18} = 262144$ possible response patterns exists. Due to relation $\preceq_{\delta_1}$, 175 of the patterns are states (i. e., 0.067 % of all possible patterns). Weakened relations are also considered. In Table 5 the cardinalities of the corresponding quasi-ordinal knowledge spaces $\mathcal{K}_1$, $\mathcal{K}_2$, and $\mathcal{K}_3$ are presented.

Model 2

In contrast to Model 1, the assumptions concerning the structure imposed on the attributes in component d are modified. The following considerations lead

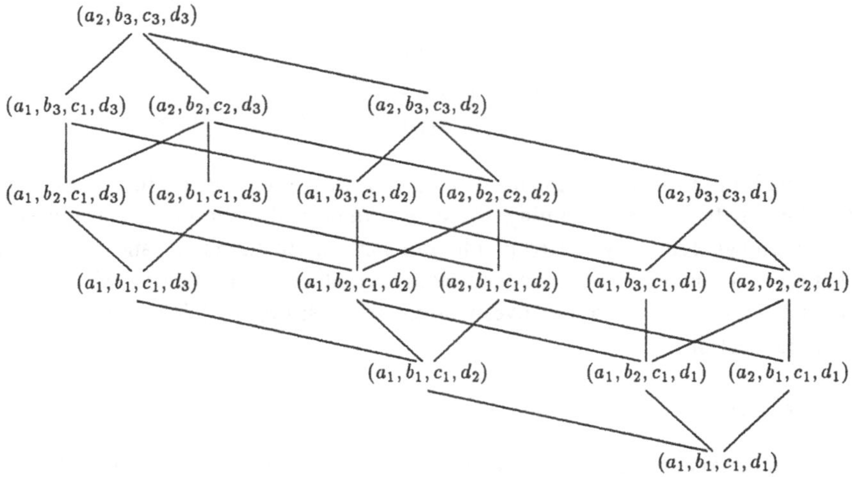

FIG. 3. Componentwise relation $\preceq_{\delta_1}$ defined on the body of 18 problems

TABLE 5
Model 1: Numbers of States in the Quasi-Ordinal Knowledge Spaces $\mathcal{K}_1, \mathcal{K}_2, \mathcal{K}_3$ Corresponding to Relations $\preceq_{\delta_1}, \preceq_{\delta_2}, \preceq_{\delta_3}$

Relation/Space	*Number of Corresponding States*	*% of Possible Patterns*
$\preceq_{\delta_1}/\mathcal{K}_1$	175	0.067
$\preceq_{\delta_2}/\mathcal{K}_2$	738	0.282
$\preceq_{\delta_3}/\mathcal{K}_3$	4136	1.578

to the model. The examination of coherence and cohesion of the problem texts yields that the textual version related to attribute d_3 is less coherent and cohesive than the versions related to attributes d_1 and d_2. For this model, the opinion that it cannot be decided which of versions 1 and 2 is more coherent and cohesive is supported. In addition, it is supposed that either comprehending version 1, or version 2, or both of the versions is a prerequisite for comprehending version 3. These assumptions are based on the consideration that it depends on the *individual* way of processing a text, whether the more compact and concise text of version 2, or the more elaborate and "wordy" text of version 1 allows easier comprehension.[10] The following demand multiassignment is assumed to hold:

$$\begin{aligned} \omega_d(d_1) &= \{\{\mathcal{O}_{11}\}\}, \\ \omega_d(d_2) &= \{\{\mathcal{O}_{12}\}\}, \\ \omega_d(d_3) &= \{\{\mathcal{O}_{11}, \mathcal{O}_{13}\}, \{\mathcal{O}_{12}, \mathcal{O}_{13}\}\}. \end{aligned}$$

The corresponding attribute mapping is derived as proposed above:

$$\begin{aligned} \varphi(d_1) &= \{\{d_1\}\}, \\ \varphi(d_2) &= \{\{d_2\}\}, \\ \varphi(d_3) &= \{\{d_1, d_3\}, \{d_2, d_3\}\}. \end{aligned}$$

The resulting surmise function σ_{δ_1} (i. e., the δ_1-weakened surmise function) leads to a knowledge space containing 479 states. Table 6 also shows the cardinalities of the knowledge spaces corresponding to surmise functions σ_{δ_2} and σ_{δ_3}. Note that $\mathcal{K}_3$, $\mathcal{K}_4$ and $\mathcal{K}_5$ are knowledge spaces (i.e., closed under union), whereas all other spaces proposed in this investigation are quasi-ordinal (closed under union and intersection).

Model 3

Another assumption to be tested is that the textual variations have no influence on solution performance. Consequently, problems only differing with respect to

[10] For a detailed discussion of the theoretical background leading to this hypothesis see Held (1993).

TABLE 6
Model 2: Numbers of States in the Knowledge Spaces
$\mathcal{K}_4, \mathcal{K}_5, \mathcal{K}_6$ Corresponding to Surmise Functions $\sigma_{\delta_1}, \sigma_{\delta_2}, \sigma_{\delta_3}$

Surmise Function/Space	*Number of Corresponding States*	*% of Possible Patterns*
$\sigma_{\delta_1}/\mathcal{K}_4$	479	0.183
$\sigma_{\delta_2}/\mathcal{K}_5$	1576	0.601
$\sigma_{\delta_3}/\mathcal{K}_6$	9950	3.800

TABLE 7
Model 3: Numbers of States in the Quasi-Ordinal
Knowledge Spaces $\mathcal{K}_{10}, \mathcal{K}_{11}, \mathcal{K}_{12}$ Corresponding to Relations $\preceq_{\delta_1}, \preceq_{\delta_2}, \preceq_{\delta_3}$

Relation/Space	*Number of Corresponding States*	*% of Possible Patterns*
$\preceq_{\delta_1}/\mathcal{K}_7$	10	0.004
$\preceq_{\delta_2}/\mathcal{K}_8$	15	0.006
$\preceq_{\delta_3}/\mathcal{K}_9$	26	0.010

component d are treated as being equivalent. This means that the problems associated with attribute tuples (a_1, b_1, c_1, d_1), (a_1, b_1, c_1, d_2), and (a_1, b_1, c_1, d_3) become elements of a notion. Altogether, six notions are obtained. If the demand assignments for components a, b and c are equal to those of Model 1, the resulting quasi-ordinal space (corresponding to relation $\preceq_{\delta_1}$) consists of 10 states. Consider Table 7 for an overview.

Method

In this investigation, 18 problems were presented to the subjects. The setting was fully computerized.[11]

Subjects

Altogether 62 subjects took part in the investigation. With 10 of them, preliminary versions of the experimental procedure were tested. Therefore, their data are not used for analysis. Two of the remaining 52 subjects did not finish the experimental run because they had difficulties in coping with the experimental task. Hence, the results of 50 subjects (14 female, 36 male) are considered. Their ages ranged from 17 to 36 years (average: 25.5 years; the lower age limit was set to 14 years). 46 of the subjects have a university entrance qualification (i. e., the "Abitur" in Germany). A total of 33 subjects had taken one or more stochas-

[11] Thanks to Dorothee Klein who was the experimenter in this investigation.

tics courses at high-school (the German "Gymnasium"). 44 subjects are students enrolled at the University of Heidelberg.

Apparatus

The experiment was run on an IBM-compatible personal computer (80386 processor), equipped with a VGA graphics card and a 14 inch color monitor presenting 9 × 16 dot characters with a refresh rate of 70 Hz. The screen intensity was adjusted to an easy reading level and was maintained at that level throughout the experiment. A second monitor was connected to the computer in order to control the subjects' actions. As input device, a standard AT keyboard with German layout was used. The experimental program was created by means of the software package "MEL" (Micro Experimental Laboratory).[12]

Procedure

The experimental sessions took place in a laboratory of the Psychological Institute of the University of Heidelberg. The experimental procedure was fully computerized. Instruction texts and problems were presented on a computer monitor, the subjects were required to enter responses via a computer keyboard. All relevant data were stored automatically.

Because the investigation is dealing with comprehension of problem texts, the presentation of problems was designed in a way that has proven to be adequate for several text comprehension paradigms (see, e. g., Graesser & Riha, 1984). The procedure was structured as follows: The subjects had to read each problem sentence by sentence. The maximum time for reading a sentence was limited to 10 seconds. The subjects could, however, proceed to the next sentence before the 10 seconds were over (by pressing the "+"-key). Hence, text presentation was designed as a mixture of a reading task with fixed reading times, and a "subject-paced" reading task. There was no possibility for scrolling back to a previous sentence. The subjects were told that they should only proceed to the next sentence if they were sure that they had "fully comprehended" the sentence displayed. The reading time for each sentence was stored. The subjects were not permitted to make written notes while reading the problems.

This type of problem presentation was used because it is assumed that differences in processing the different types of problem texts are more likely to occur if reading time is limited. With an unlimited reading time, the subjects would in any case have had enough time to perform, for instance, the reorganizations necessary for transforming texts with wording of version 3 into coherent text-bases.

As soon as the complete problem text had been read, the subject was asked (by a message on the screen) to start working on the problem. The subjects were required to write down all computations and the solution on a separate sheet of

[12] For a description and detailed information concerning MEL see Schneider (1988).

paper. An intermediate message was shown for 60 seconds, then the "response screen" was displayed.[13] This screen contained a "response field", on which the cursor was positioned. At the bottom of the response screen, the time remaining for working on the problems was displayed. Starting with the presentation of the response screen, the subject had 7 minutes for computing the result and typing it into the response field. If an answer had been given, the subject was required to press a function key. Then the presentation of the next problem was started.

The problems were presented in a fixed sequence. First, two "warming-up" problems had to be solved. These problems were not considered for the results. The 18 problems relevant for the investigation were—all in all—presented in order of increasing hypothetical difficulty (as concerns components a, b, and c). With respect to textual versions, version 3 of a problem (based on attribute d_3) was always presented first, then came version 2 (based on attribute d_2), and then version 1 (based on attribute d_1). This was done due to the consideration that fatigue or motivational decrease would have more influence on solution performance for the problems with less complex wording.

Results

In Table 8, the response patterns obtained in the investigation are shown. A subject's response is coded as "0", if it does not match one of the possible solutions, whereas it is coded as "1" if a valid solution is observed. Two of the subjects could answer each of the problems correctly, while two subjects could not answer any of the problems. Altogether, 37 different response patterns occurred. 564 of 900 answers (= 50×18) are correct (62.7 %), 336 are incorrect (37.3 %).

As next step, we show how well those data agree with Models 1 to 3 proposed above. In Table 9, the frequencies of patterns congruent with the hypotheses, and the symmetric distances between noncongruent patterns and the "closest" states are presented. Concerning Model 1 (the "basic" model), 21 of the subjects generated a data pattern being congruent with a state in $\mathcal{K}_1$. We have to consider that $\mathcal{K}_1$ only contains 0.067 % of the possible patterns. Space $\mathcal{K}_3$ corresponding to the δ_3-weakened surmise relation, fits the data "better." This was expected as $\mathcal{K}_1$ is a proper subset of $\mathcal{K}_3$. Although $\mathcal{K}_3$ is about 24 times as large as $\mathcal{K}_1$, we have to be aware that $\mathcal{K}_3$ contains not more than 1.578 % of the "potential" data patterns. Furthermore, the patterns of 12 subjects do have a symmetric distance of 1 to the "closest" state in $\mathcal{K}_3$. One can also see that there are, with respect to $\mathcal{K}_3$, 25 different congruent patterns. This indicates (as could already be seen in Table 8) that no cumulations of—possibly trivial[14]—congruent patterns occurred.

The spaces of Model 2 correspond to a surmise mapping that is based on a de-

[13] This intermediate message was set in order to avoid subject responses from being entered too quickly. The intention was to ensure that each subject took at least 60 seconds for thinking about the solution.

[14] Note that both, $\{\emptyset\}$, and $\{Q\}$ (with Q the problem set) are states.

TABLE 8
Response Patterns Obtained in the Investigation (Problems 1,. . . ,18)

Freq.	*Pattern*	*Freq.*	*Pattern*
5	111111111110100000	1	111111101111101110
3	101010000000000000	1	111111101110101011
2	111111111111111111	1	111110111111100011
2	111111111111111110	1	111110111110000000
2	111111111111110111	1	111011111110111001
2	111111111111001111	1	111011111110101001
2	111011001000000000	1	111011111110100000
2	101111111110100000	1	101111111110001011
2	000000000000000000	1	101111011110100000
1	111111111111111010	1	101110111110000000
1	111111111111101111	1	101110110110100000
1	111111111111101011	1	101011001000000000
1	111111111111101000	1	101011000000000000
1	111111111110111010	1	100111111010100000
1	111111111110101001	1	011101001100100000
1	111111111110001000	1	011011011100000000
1	111111111100111011	1	001111101111111111
1	111111111010100000	1	000100000000000000
1	111111110101101111		

Total frequencies of correct solutions: Problem 1: 44 subjects; P2: 34; P3: 46; P4: 37; P5: 46; P6: 40; P7: 37; P8: 36; P9: 41; P10: 38; P11: 36; P12: 16; P13: 33; P14: 11; P15: 21; P16: 12; P17: 19; P18: 17

mand multiassignment. The response patterns of 31 subjects agree with states in $\mathcal{K}_4$ (derived from a δ_1-weakened surmise function). 12 patterns have a symmetric distance of 1 ($\overline{d} = 0.62$). Comparing space $\mathcal{K}_5$ of this model (34 congruent patterns; $\overline{d} = 0.40$; 1576 states) with space $\mathcal{K}_3$ of Model 1 (36 congruent patterns; $\overline{d} = 0.32$; 4136 states) it can be seen that $\mathcal{K}_5$ seems to explain the data nearly as well as $\mathcal{K}_3$, that contains two and a half times as many states. $\mathcal{K}_6$, consisting of 9950 states (3.8 % of the possible patterns), obviously fits the data rather well—43 patterns out of 50 are congruent with a state.

For Model 3, the assumption has been made that problems whose attribute tuples only differ in the attribute of component *d* are equivalent. Evidently, the data fit the spaces worse than those of the other two models.

Surely, each of the (deterministic) hypotheses has—in principle—to be rejected, because noncongruent solution patterns occurred. However it is to be noted that especially "careless errors" may have influenced the results. If the probability of a careless error is set to 0.05, the probability for obtaining a pattern that is congruent to a subject's knowledge state is equal to $0.95^{18} \approx 0.397$. The relative frequencies of congruent patterns in the data, however, are, as concerns Models 1 and 2, higher than this probability.

Discussion

All in all, the results of this investigation are quite satisfying; neither a floor-, nor a ceiling-effect could be observed. 50 subjects produced 37 different response patterns. If we regard space $\mathcal{K}_1$ containing a total of 175 states, the result of 21 congruent patterns and 18 patterns with a symmetric distance of 1 strongly indicates the validity of the hypothesis. Although the δ_3-weakened structure of Model 1 (space $\mathcal{K}_3$) still can be viewed as strong assumption (1.578 % of 262144 possible patterns are states), the corresponding result is outstanding: the patterns of only two subjects differ in more than one element from a state in $\mathcal{K}_3$.

It appears, however, that through the demand multiassignment of Model 2, the data are described even better. In addition, the results corresponding to Model 4 indicate that the variations in wording do have an influence on solution performance. The assumption of this model, namely that problems only differing in wording are equivalent, can clearly be rejected.

In general, in each of the models, assumptions concerning an influence of

TABLE 9
Results of the Investigation

Knowledge space ($\mid\mathcal{K}_n\mid$)	*Number of patterns with a symmetric distance of* 0	1	2	3	4	5	$\overline{d}$	*Number of diff. congruent patterns*	*Number of diff. non congruent patterns*
					Model 1				
$\mathcal{K}_1$ (175)	21	18	7	3	1	–	0.90	14	25
$\mathcal{K}_2$ (738)	25	19	6	–	–	–	0.62	15	22
$\mathcal{K}_3$ (4136)	36	12	2	–	–	–	0.32	25	12
					Model 2				
$\mathcal{K}_4$ (479)	31	12	3	3	1	–	0.62	20	17
$\mathcal{K}_5$ (1576)	34	12	4	–	–	–	0.40	23	14
$\mathcal{K}_6$ (9950)	43	5	2	–	–	–	0.18	31	6
					Model 3				
$\mathcal{K}_7$ (10)	15	15	9	8	2	1	1.40	6	31
$\mathcal{K}_8$ (15)	15	15	10	7	3	–	1.36	6	31
$\mathcal{K}_9$ (26)	15	15	10	9	1	–	1.32	6	31

Congruent states have a symmetric distance of 0.
Column '$\overline{d}$' contains the mean distances.

wording on solution performance have been expressed. Comparing those models, we see that the attribute assignment of Model 2 occurs to be the most adequate approach (out of 3 models) for predicting the data. Furthermore, ignoring textual variations leads to particularly poor results (cf. Model 3).

It is clear that with increasing size of a knowledge space, the number of congruent response patterns will also grow. Therefore the size of a knowledge space must be kept in mind when the goodness of fit with respect to a set of data is considered.

Due to the lack of a probabilistic model, as well as statistical criterions for determining how good the data fit the hypotheses and for deciding which of the hypotheses the data fit best, each of the statements made above is of a more or less interpretative nature. Note that the dataset is too small for performing parameter estimations necessary for applying standard methods of inferential statistics.

GENERAL DISCUSSION

It is characteristic of the research presented in this work that various different theoretical approaches are integrated into the framework of psychological knowledge assessment. The considerations are in several respects influenced by the theory of knowledge spaces (Doignon & Falmagne, 1985, 1998), which provides elaborate concepts for the formal representation of knowledge domains. The properties of diagnostic problems that are one of the central topic of this chapter are, however, not subject matter of knowledge space theory. The presented considerations and results concern mainly an integrated concept for problem generation, problem coding, and the component-based derivation of problem structures. It is clarified how diagnostic problems can be generated in a quasi-automatic manner, how problem coding is related to the generative process, and how various formal principles for establishing problem structures can be applied to sets of unambiguously coded problems.

How this integrated concept can be utilized "in practice" is demonstrated for the domain of elementary stochastics. The proposed formal apparatus is, however, sufficiently general in order to be applicable to arbitrary knowledge domains. Obviously, the use of generative grammars is particularly adequate for the construction of word problems. However, in principle, every type of problem can be generated by means of grammars.

It is stressed that rule-based problem generation is a necessary prerequisite for unambiguous problem coding and for the definition of a class of possible problem variations. A subclass of those variations is related to varying problem demands.

The general principles for deriving problem structures depend both on the existence of problem codings and on theoretically based demand assignments. Through demand assignments, a psychological foundation of the problem structures is achieved. In contrast to related approaches that assign demands or "skills" to problems as a whole, the approach proposed here facilitates to determine which variations of problem formulation entails a variation of demands on the problem-solver.

The methods for a systematic weakening of knowledge structures (δ_n-weakened surmise relations and σ_n-weakened surmise functions) are introduced in order to achieve a higher "flexibility" in stating experimental hypotheses. According to these methods, assumptions concerning demands associated with problem components can be dropped in a systematic way.

The principle of rule-based problem construction also facilitates the investigation of problem components that are not directly related to the focused knowledge domain. This is demonstrated by using the concepts of textual coherence and textual cohesion for the definition of a problem component. It is one of the most important achievements of this work that it could be shown that the quite different approaches of Kintsch and van Dijk (1978) and Brainerd (1977) can be integrated in order to establish component-based problem structures.

In general, the results of the empirical investigations confirm the functionality of this problem oriented approach in an impressive manner. Obviously, a precise problem analysis together with a clear differentiation between problem formulation and the related demands on the problem-solver is an important contribution to the efficiency of knowledge assessment.

ACKNOWLEDGEMENTS

The research reported in this paper was supported by Grant Lu 385/1 of the Deutsche Forschungsgemeinschaft to J. Lukas and D. Albert at the University of Heidelberg.

REFERENCES

Albert, D. (1989). *Knowledge assessment: Choice heuristics as strategies for constructing questions and problems*, Paper presented at the 20th European Mathematical Psychology Group Meeting, Nijmegen, September 1989.

Albert, D., & Held, T. (1994). Establishing knowledge spaces by systematical problem construction. In D. Albert (Ed.), *Knowledge structures* (pp. 81–115). Berlin: Springer.

Bart, W. M., & Krus, D. J. (1973). An ordering-theoretic method to determine hierarchies among items. *Educational and Psychological Measurement*, *33*, 291–300.

Birkhoff, G. (1973). *Lattice theory* (3rd ed.). Providence: American Mathematical Society.

Brainerd, B. (1977). Graphs, topology and text. *Poetics*, *6*, 1–14.

Chomsky, N. (1959). On certain formal properties of grammars. *Information and Control*, *2*, 137–167.

Doignon, J.-P. (1994). Knowledge spaces and skill assignments. In G. H. Fischer &

D. Laming (Eds.), *Contributions to mathematical psychology, psychometrics, and methodology* (pp. 111–121). Berlin: Springer.

Doignon, J.-P., & Falmagne, J.-C. (1998). *Knowledge spaces.* Berlin: Springer.

Doignon, J.-P., & Falmagne, J.-C. (1985). Spaces for the assessment of knowledge. *International Journal of Man-Machine Studies, 23,* 175–196.

Dowling, C. E. (1993). Applying the basis of a knowledge space for controlling the questioning of an expert. *Journal of Mathematical Psychology, 37,* 21–48.

Falmagne, J.-C. (1989). A latent trait theory via a stochastic learning theory for a knowledge space. *Psychometrica, 54*(2), 283–303.

Falmagne, J.-C., & Doignon, J.-P. (1988a). A class of stochastic procedures for the assessment of knowledge. *British Journal of Mathematical and Statistical Psychology, 41,* 1–23.

Falmagne, J.-C., & Doignon, J.-P. (1988b). A markovian procedure for assessing the state of a system. *Journal of Mathematical Psychology, 32*(3), 232–258.

Falmagne, J.-C., Koppen, M., Villano, M., Doignon, J.-P., & Johannesen, L. (1990). Introduction to knowledge spaces: how to build, test and search them. *Psychological Review, 97,* 201–224.

Graesser, A. C., & Riha, J. R. (1984). An application of multiple regression techniques to sentence reading times. In D. E. Kieras & M. A. Just (Eds.), *New methods in reading comprehension research* (pp. 183–218). Hillsdale, NJ: Lawrence Erlbaum Association.

Guttman, L. A. (1957). *Introduction to facet design and analysis.* Proceedings of the fifteenth international congress of psychology, Brussels. Amsterdam: North-Holland.

Held, T. (1992). *Systematische Konstruktion und Ordnung von Aufgabenmengen zur elementaren Wahrscheinlichkeitsrechnung* [Systematical construction and ordering of problem sets of elementary stochastics], Paper read at the 34th Tagung experimentell arbeitender Psychologen, Osnabrück, 12.-16. April 1992.

Held, T. (1993). *Establishment and empirical validation of problem structures based on domain specific skills and textual properties—A contribution to the "Theory of Knowledge Spaces".* Unpublished doctoral dissertation, University of Heidelberg, Germany.

Herbst, K. (1978). *Ermittlung und Bewertung von Verstößen gegen den Grundsatz der spezifischen Objektivität in psychodiagnostischen Untersuchungen* [Determination and assessment of violations of specific objectivity in psychodiagnostical investigations], Dissertation, Universität Regensburg.

Kambouri, M., Koppen, M., Villano, M., & Falmagne, J.-C. (1994). Knowledge assessment: Tapping human expertise by the QUERY routine. *International Journal of Human-Computer Studies, 40,* 119–151.

Kintsch, W. (1974). *The representation of meaning in memory.* Hillsdale: Lawrence Erlbaum Association.

Kintsch, W., & van Dijk, T. A. (1978). Toward a model of text comprehension and production. *Psychological Review, 86,* 363–394.

Koppen, M. (1993). Extracting human expertise for constructing knowledge spaces: An algorithm. *Journal of Mathematical Psychology, 37,* 1–20.

Lipski, J. M. (1974). A topology of semantic dependence. *Semiotica, 12,* 145–170.

Lukas, J. (1991). *Knowledge structures and information systems,* Paper presented at the 22nd European Mathematical Psychology Group Meeting, Vienna, September 1991.

Marcus, S. (1980). Textual cohesion and textual coherence. *Revue Romain de Linguistique, XXV*(2), 101–112.

Scandura, J. M. (1973). *Structural learning I. Theory and research.* New York: Gordon & Breach.

Scheiblechner, H. (1972). Das Lernen und Lösen komplexer Denkaufgaben. [The learning and solving of complex thinking problems.] *Zeitschrift für experimentelle und angewandte Psychologie, 24*, 477–506.

Schneider, W. (1988). Micro Experimental Laboratory: An integrated system for IBM PC compatibles. *Behavior Research Methods, Instruments, & Computers, 20*(2), 206–217.

Turner, A., & Greene, E. (1977). *Constructions and use of a propositional text base*, Technical Report 63, Colorado University.

Van Buggenhaut, J., & Degreef, E. (1987). On dichotomisation methods in boolean analysis of questionaires. In E. E. Roskam & R. Suck (Eds.), *Progress in Mathematical Psychology* (pp. 447–453). Amsterdam: Elsevier.

Van Leeuwe, J. F. J. (1990). *Probabilistic conjunctive models—Contributions to multidimensional analysis of binary test Data*, NICI Technical Report 90-11, Nijmegen Institute for Cognition Research and Information Technology.

Villano, M. (1991). *Computerized Knowledge Assessment: Building the Knowledge Structure and Calibrating the Assessment Routine*, PhD thesis, New York University.

5

Modeling Knowledge as Competence and Performance

Klaus Korossy
Ruprecht-Karls-Universität Heidelberg

This chapter develops an extension of Doignon and Falmagne's knowledge structures theory by integrating it into a *competence-performance conception*. The aim is to show one possible way in which the purely behavioral and descriptive knowledge structures approach could be structurally enriched in order to account for the need of explanatory features for the empirically observed solution behavior. *Performance* is conceived as the observable solution behavior of a person on a set of domain-specific *problems*. *Competence (ability, skills)* is understood as a theoretical construct accounting for the performance. The basic concept is a mathematical structure termed a *diagnostic*, that creates a correspondence between the competence and the performance level. The concept of a *union-stable diagnostic* is defined as an elaboration of Doignon and Falmagne's concept of a knowledge space. Conditions for the construction and several properties of union-stable diagnostics are presented. Finally, an empirical application of the competence-performance conception in a small knowledge domain is reported that shall illustrate some advantages of the introduced modeling approach.

THE KNOWLEDGE STRUCTURES APPROACH

The knowledge structures theory (Doignon & Falmagne, 1985, 1998; see also Falmagne, Koppen, Villano, Johannesen & Doignon, 1990) presupposes that the "knowledge" of an individual in a particular domain of knowledge can be operationalized as the solving behavior of that individual on a domain-specific set X of *problems*. If the solution result for each problem is binarily coded by correct/incorrect, then the *knowledge state* of an individual in the given field of knowledge can be formally described as the subset of problems from X he or she is capable of solving. The well-known fact that solution dependencies often exist between problems of a certain field of knowledge is taken into account by two distinct approaches.

The first approach is obvious. If solution dependencies exist on the set X of problems, then not each subset of X will indicate an expectable solution pattern; instead, the set of all "empirically expectable" solution patterns constitutes a particular family K of subsets of X that is not identical with the power set of X. Formally, the pair (X, K) is said to be a *knowledge structure*, the elements of (X, K) are called *knowledge states*. Special interest is directed to *knowledge spaces*. A knowledge structure (X, K) is called a *knowledge space* when $\emptyset$, $X \in K$ and K is stable under union. Arguments for this conceptualization can be found in Doignon (1994a), in Doignon and Falmagne (1985), or in Falmagne et al. (1990). One reason for the importance of knowledge spaces is that they can be economically stored in the form of a *basis*. The *basis* of a knowledge space (X, K) is the minimal subfamily $\mathcal{B}(K)$ of K so that each knowledge state in K can be written as a union of elements of $\mathcal{B}(K)$.

The second approach for modeling the solution dependencies among different problems is realized by the concept of a *surmise system*. The idea is to associate each problem $x \in X$ with a family of subsets of X called *clauses*, with the interpretation that, if a person is capable of solving x then he or she is capable of solving all problems in at least one of these clauses. Formally, a *surmise system* is defined as a pair (X, σ) with a set X of problems and a function σ, called *surmise function on* X, that assigns to each $x \in X$ a family $\sigma(x)$ of subsets of X, the *clauses* for x, so that the following three postulates are satisfied:

(1) Any clause for x contains x.
(2) If some clause C for x contains some problem $y \in X$, then there exists some clause D for y satisfying $D \subseteq C$.
(3) Any clause C for x is minimal with respect to $\subseteq$.

The two concepts of a *knowledge space* and of a *surmise system* play a central role in Doignon and Falmagne's theory. In fact, Doignon and Falmagne (1985) succeeded in constructing a one-to-one correspondence between surmise systems and knowledge spaces. For a given knowledge space (X, K), let K_x denote the family of states containing some problem $x \in X$, and let $\hat{K}_x := \text{Min } K_x$ denote the set of all minimal states in K_x (with respect to the subset relation $\subseteq$). Then, the following theorem can be proved.

THEOREM (Doignon & Falmagne, 1985): Let X be a non-empty, finite set of problems. Then there exists a one-to-one correspondence between the set of all surmise systems (X, σ) on X and the set of all knowledge spaces (X, K) on X. This correspondence is specified by the equation

$$\sigma(x) = \hat{K}_x \ (x \in X).$$

REMARK 5.1 To supply the statement in the previous theorem by a relation later used several times, the families $\hat{K}_x$ $(x \in X)$ are closely related to the basis $\mathcal{B}(K)$ of the knowledge space through $\mathcal{B}(K) = \bigcup\{\hat{K}_x \mid x \in X\}$.

Knowledge spaces and surmise systems capture those structures inherent in knowledge that can be operationalized as solvability dependencies among problems of a specified field of knowledge. Both concepts can be applied to represent knowledge under different viewpoints but in formally equivalent ways.

In Doignon and Falmagne's set-theoretical approach to knowledge representation, knowledge states are modeled as empirically expectable solution patterns on problems. In this way, a knowledge representation model can be directly exposed to empirical validation. Once the knowledge model has passed the validational examination, it can be immediately utilized as a frame for individual knowledge diagnosing. Moreover, automatic procedures for an efficient assessment of knowledge can be designed (see Doignon, 1994a; Doignon & Falmagne, 1985; Falmagne & Doignon, 1988a, 1988b). The efficient and economic assessment of knowledge is the essential aim and, without doubt, the decisive advantage of the knowledge structures theory.

Unfortunately, a serious deficiency of the knowledge structures theory is its inherent empirical and descriptive character. The most critical point of this theory in practical application concerns the establishment of the family of knowledge states by interviewing experts. Sophisticated methods utilizing another equivalent to knowledge spaces have been developed in order to optimize the querying procedures (see, e.g., Dowling, 1993, 1994; Koppen & Doignon, 1990). Nevertheless, the knowledge spaces are empirically established structures without any theoretical foundation. An approach for basing a knowledge structure on a family of "skills" is outlined in Falmagne et al. (1990), and developed further in Doignon (1994b) and recently in Düntsch and Gediga (1995). But, in this line of research, skills are treated as pure epiphenomena of empirically constituted structures, as posthoc introduced "hidden factors" (Doignon, 1994b) that formally generate, but do not explain the previously established knowledge structure. However, features for founding knowledge states on an *explanatory meaningful (domain-specific) theory* seem essential with respect to practical application for at least two reasons:

First, only from such a meaningful theory can hypotheses concerning the solution behavior on problems of a new sample be derived. Second, only a meaningful theory that explains solution behavior can give an indication of which information a person should be taught in order to enable him or her to master a previously unsolved problem.

In contrast to the basically behavioral features of the knowledge structures approach, several approaches for the *constructional* genesis of knowledge structures have been developed during the last years. They have in common that they propose methods for *constructing* knowledge structures by a systematical analysis of the "basic components" of the domain-specific problems (see, e.g., Albert & Held, 1994; Held, 1993), or by relating behavioral knowledge states to existing domain-specific theorizing (Korossy, 1993; Schrepp, 1993; for a review of several approaches see Lukas & Albert, 1993).

The aim of this chapter is to introduce an extension of Doignon and Falmagne's theory within a *competence-performance conception* (Korossy, 1993). The main purpose of this conception is to provide additional features that allow for utilizing domain-specific theories for the knowledge modeling on a theoretical (competence) level; "knowledge structures" in the sense of Doignon and Falmagne's theory are then conceptualized and established as empirical (performance) representations of the theoretically constituted structures.

The next three sections develop selected parts of the competence-performance approach to the knowledge structures theory. In the final section, an empirical application of the introduced theory within a domain of elementary arithmetic is reported. Altogether, this contribution outlines one possible approach in which Doignon and Falmagne's knowledge structures theory could be reconciled with long-standing psychometrical traditions.

THE COMPETENCE-PERFORMANCE APPROACH

The basic premise is to make a clear distinction between *competence (skills, ability)* and *performance*. *Performance* refers to the empirically observable solution behavior on certain given problems. *Competence* is conceptualized in terms of theoretically founded entities accounting for the observable solution behavior. The theoretical conception outlined in this chapter is based on the following assumptions.

ASSUMPTIONS 5.1 Given a specific knowledge domain $\mathcal{W}$ and a target population $\mathcal{N}$.

(1) The knowledge of the target population $\mathcal{N}$ relative to the domain $\mathcal{W}$ can be modeled through a finite, non-empty family $\mathcal{K}$ of "*competence states*". The *knowledge, ability,* or the *set of skills* of a person from $\mathcal{N}$ with respect to

the domain $\mathcal{W}$ (at a given time) is a certain element of $\mathcal{K}$. The competence states are conceived as not directly observable theoretical constructs that (ideally) are constituted by the support of some domain-specific theory.

(2) The family $\mathcal{K}$ of competence states may be structured through order relations or algebraic operations. Taken into special consideration is the case that the competence states themselves are specified as particular subsets of a domain-specific set $\mathcal{E}$ of "*elementary competencies*". The set $\mathcal{E}$ may be structured for its part (e.g., through surmise structures); these structures would then limit the family $\mathcal{K}$ of competence states.

(3) The knowledge of a person from $\mathcal{N}$ relative to the domain $\mathcal{W}$ can also be modeled by means of the solution behavior on a set $\mathcal{A}$ of domain-specific "*problems*". It is assumed that

 (a) each problem $x \in \mathcal{A}$ is solvable exclusively with the knowledge modeled by $\mathcal{K}$;
 (b) for each problem $x \in \mathcal{A}$ and each competence state $\kappa \in \mathcal{K}$ it is uniquely determined whether or not x can be solved in κ.

It is explicitly taken into account that a problem may be solvable in various ways (in different competence states respectively).

(4) Every person from $\mathcal{N}$ is, according to his or her momentary competence state, capable of solving certain problems of $\mathcal{A}$ and only those problems. The result of this solving behavior is observable as a "*correct solution*" or an "*incorrect solution*" for each applied problem. The subset of correctly solved problems is called the "*solution pattern*" of that person.

(5) The solvability conditions of the problems of $\mathcal{A}$ in $\mathcal{K}$ provide theoretically founded hypotheses on empirically expectable solution patterns. A theoretically expected solution pattern is called a "*performance state*".

(6) An empirical validation of the knowledge modeling means that the performance states coincide with the observed solution patterns. If the modeling is validated, the performance state of a person suggests conclusions on the possible competence state of that person.

As a starting point for developing these assumptions and integrating the concepts of the knowledge structures theory in the competence-performance approach, we reinterpret the basic concepts of the knowledge structures theory in the following manner:

- A pair $(\mathcal{E}, \mathcal{K})$ consisting of a finite, non-empty set $\mathcal{E}$ whose elements are called *elementary competencies*, and a non-empty family $\mathcal{K}$ of subsets of $\mathcal{E}$ called *competence states*, is said to be a *competence structure*. Hereby it is assumed that for each $\varepsilon \in \mathcal{E}$ there exists a competence state $\kappa \in \mathcal{K}$ such that $\varepsilon \in \kappa$.[1]

[1] This postulate is equivalent to $\cup\mathcal{K} = \mathcal{E}$, so that $\mathcal{K}$ itself contains all information about $\mathcal{E}$.

If $\emptyset, \mathcal{E} \in \mathcal{K}$ and $\mathcal{K}$ is stable under union, then $(\mathcal{E}, \mathcal{K})$ is called a *competence space*. Often $(\mathcal{E}, \mathcal{K})$ is denoted by $(\mathcal{K}, \cup)$.

- A pair $(\mathcal{A}, \mathcal{P})$ consisting of a finite, non-empty set $\mathcal{A}$ of *problems* and a non-empty family $\mathcal{P}$ of subsets of $\mathcal{A}$ called *performance states*, is said to be a *performance structure*. Again, it is assumed that for each $x \in \mathcal{A}$ there exists a performance state $Z \in \mathcal{P}$ such that $x \in Z$.

  If $\emptyset, \mathcal{A} \in \mathcal{P}$ and $\mathcal{P}$ is stable under union, then $(\mathcal{A}, \mathcal{P})$ is called a *performance space*. The short denotation is $(\mathcal{P}, \cup)$.

Up until this point, competence and performance structures and competence and performance spaces represent formally equivalent structures that are only distinguished by their interpretational context. However, a central task of the competence-performance conception, demanded by the Assumptions 5.1, is to create explicitly defined relations between the competence and the performance level. This task is initiated by introducing the basic concept of a *diagnostic*.

THE CONCEPT OF A DIAGNOSTIC

The concept of a *diagnostic* is developed in two steps. The first step introduces the concept of a competence-based problem set; the second step leads directly to the concept of a diagnostic.

The concept of a *competence-based problem set* involves the following idea that is suggested by the Assumptions 5.1: A problem $x \in \mathcal{A}$ can be related to the set $\mathcal{K}$ of competence states or *"interpreted" in* $\mathcal{K}$, if $\mathcal{K}$ includes all competence states of the domain—each of which enables solving that problem. Under that condition, the *interpretation* of x in $\mathcal{K}$ can be realized through mapping x to a problem-specific set

$$k(x) := \{\kappa_1, \kappa_2, \kappa_3, \ldots, \kappa_r\} \quad (r \in \mathbb{N}),$$

which contains exactly those competence states of $\mathcal{K}$, in which x can be solved. This conceptualization especially accounts for the well-known fact that a problem can often be solved using different ways (see Assumption 5.1 (3)). This consideration motivates the following definition.

DEFINITION 5.1 Let $\mathcal{K}$ be a set of competence states. Furthermore, let $\mathcal{A}$ be a set of problems and $k : \mathcal{A} \longrightarrow \wp(\mathcal{K})$ a function that assigns to each problem $x \in \mathcal{A}$ a subset $k_x := k(x) \subseteq \mathcal{K}$ of competence states (the set of competence states in each of which x is solvable) so that

(k1) $k_x \neq \emptyset$;
(k2) $k_x \neq \mathcal{K}$.

Then the problem set $\mathcal{A}$ is called a problem set *based on* $\mathcal{K}$ or *interpreted in* $\mathcal{K}$; k is called the *interpretation function*, and, for each problem $x \in \mathcal{A}$ the set k_x is called the *interpretation of* x *in* $\mathcal{K}$.

Definition 5.1 formalizes the idea of a *competence-based problem set*. The existence (well-definedness) of the interpretation function requires that to each problem a unique interpretation *within* the considered set of competence states is assigned. The postulates $k(\mathcal{A}) \subseteq \wp(\mathcal{K})$ and (k1) mean that no "*practically relevant*" solving competencies should exist *outside* $\mathcal{K}$. The postulate $k_x \neq \mathcal{K}$, for each problem $x \in \mathcal{A}$, additionally accounts for the idea that solving or not solving a problem seen as representative for the domain should involve some information on the underlying competence states.

The second step towards the concept of a diagnostic requires a reasonable concept of a *competence-based performance state*. Let $\mathcal{A}$ be a set of problems interpreted in a set of competence states $\mathcal{K}$ by an interpretation function k. Then, a subset $Z \subseteq \mathcal{A}$ of problems is considered a *performance state*, when there exists a competence state $\kappa \in \mathcal{K}$ so that Z includes exactly those problems that are solvable in κ. This idea is made concrete by the function

$$p : \begin{cases} \mathcal{K} \longrightarrow \wp(\mathcal{A}) \\ \kappa \longmapsto p(\kappa) := \{x \in \mathcal{A} \mid \kappa \in k_x\} \end{cases} \tag{1}$$

that assigns to each competence state $\kappa \in \mathcal{K}$ the unique (possibly empty) *set of all problems solvable in* κ. Then, each element in the image of $\mathcal{K}$ under p is a (*competence-based*) *performance state* as required above, and the total set of all these performance states is a "*representation*" of the given set of competence states $\mathcal{K}$, eventually as a "diminished" image of $\mathcal{K}$.

The functions k and p are appropriate tools for linking a performance structure $(\mathcal{A}, \mathcal{P})$ to a set $\mathcal{K}$ of competence states. With the next definition we arrive at the concept of a diagnostic.

DEFINITION 5.2 Let $\mathcal{K}$ be a set of competence states and $(\mathcal{A}, \mathcal{P})$ a performance structure. Assume the following conditions are satisfied:

(1) $\mathcal{A}$ is a problem set interpreted in $\mathcal{K}$ by an interpretation function $k : \mathcal{A} \longrightarrow \wp(\mathcal{K})$;

(2) for the function $p : \mathcal{K} \longrightarrow \wp(\mathcal{A})$ induced by k according to (1) is $p(\mathcal{K}) = \mathcal{P}$.

Then, $(\mathcal{A}, \mathcal{P})$ is called a *representation of* $\mathcal{K}$ (under p). The function k is called *interpretation function for* $(\mathcal{A}, \mathcal{P})$, the function p is called *representation function for* $\mathcal{K}$. The 5–tupel $(\mathcal{K}, \mathcal{A}, \mathcal{P}, k, p)$ is said to be a *diagnostic*. If the competence states of $\mathcal{K}$ are subsets of a set $\mathcal{E}$ of elementary competencies (that is, if $(\mathcal{E}, \mathcal{K})$ is a competence structure), then the diagnostic is denoted as a 6–tupel $(\mathcal{E}, \mathcal{K}, \mathcal{A}, \mathcal{P}, k, p)$.

The concepts of a diagnostic or a performance structure $(\mathcal{A}, \mathcal{P})$ as a representation of a set $\mathcal{K}$ of competence states under a representation function p reflect the obvious idea of (competence-based) *performance states* in $\mathcal{P}$. Condition (2), which postulates that the image of $\mathcal{K}$ under the representation function p be *equal* to $\mathcal{P}$, formally assures that, on the one hand, all subsets of problems generated by p are states of the performance structure $(\mathcal{A}, \mathcal{P})$ (because $p(\mathcal{K}) \subseteq \mathcal{P}$ is required); on the other hand, (2) assures that all performance states occurring in $\mathcal{P}$ are images of competence states in $\mathcal{K}$ and in that sense are theoretically interpretable (because $p(\mathcal{K}) \supseteq \mathcal{P}$ is required).

From a structural view, the concept of a diagnostic is a rather weak one. Nevertheless, several properties can be derived (for details, see Korossy, 1993). We mention the following easily provable proposition, that additionally characterizes the relation between the competence and the performance level involved in the concept of a diagnostic, and suggests a consideration on the *equivalence* of problems.

PROPOSITION 5.1 Let $(\mathcal{E}, \mathcal{K}, \mathcal{A}, \mathcal{P}, k, p)$ be a diagnostic. Furthermore, let $\mathcal{P}_x := \{Z \in \mathcal{P} \mid x \in Z\}$, and $p(k_x) := \{p(\kappa) \mid \kappa \in k_x\}$ be the image of $k_x \subseteq \mathcal{K}$ in $\mathcal{P}$ for each $x \in \mathcal{A}$. Then

(1) $\mathcal{P}_x = p(k_x)$, *for all* $x \in \mathcal{A}$,
(2) $\mathcal{P}_x = \mathcal{P}_y \iff k_x = k_y$, *for all* $x, y \in \mathcal{A}$.

Statement (1) in Proposition 5.1 means that the subfamily of all performance states containing $x \in \mathcal{A}$ is founded and completely determined by the interpretation k_x of x in $\mathcal{K}$. A careful interpretation of statement (2) provides the possibility of theoretically founding the *equivalence* of problems. In general, if two problems x, y from a set of problems are considered *equivalent*, then each solution pattern observed in a test on this problem set should contain x whenever it contains y. Now, if $(\mathcal{A}, \mathcal{P})$ is a representation of the competence structure $(\mathcal{E}, \mathcal{K})$, i.e. a component of the diagnostic $(\mathcal{E}, \mathcal{K}, \mathcal{A}, \mathcal{P}, k, p)$, then the equivalence of problems can be theoretically based by (2): Two problems x, $y \in \mathcal{A}$ are expected to occur in the same solution patterns when they occur in the same performance states of $\mathcal{P}$, that is, when $\mathcal{P}_x = \mathcal{P}_y$; by (2) of Proposition 5.1 this is the case if and only if $k_x = k_y$, that is, in the case that the interpretations of x and y in $\mathcal{K}$ are equal. This relation provides a theoretical foundation for the rational construction of equivalent problems or, said another way, equivalent versions of the same "problem type". [2] Clearly, the equivalence of problems founded on the identity of their interpretations includes an empirical hypothesis depending on a valid competence modeling and a valid interpretation of the problems in the competence modeling.

[2] This principle is utilized in several of our empirical investigations where equivalent problem sets are constructed with respect to the practical requirements of group testings.

In general, the defined concept of a diagnostic seems to fit the basic Assumptions 5.1 fairly well. Also the benefits of the proposed conceptualization are obvious: On the level of competence, a domain-specific theory can be utilized for a genuine competence modeling. Through selecting or constructing appropriate problems that are interpretable in the set of competence states, a representing performance structure can be established. Whenever a performance structure has been constructed as a representation of a well-founded domain-specific competence structure, then the performance states can be theoretically explained and incorporate theoretically founded hypotheses on the expectable solution patterns on the problem set. In this way, a theoretical modeling in the form of a competence structure can be exposed to empirical testing and validation. When a diagnostic has been accepted as psychologically valid, it can be used for individual competence assessment, because the empirically observed solution pattern of a person will provide indication of the person's solution behavior possibly underlying competence states.

Despite several interesting aspects, from a mathematical viewpoint, the above defined concept of a diagnostic is a rather weak concept. In the following section, this concept is structurally enriched in order to redefine by it Doignon and Falmagne's concept of a knowledge space within the competence-performance approach.

UNION-STABLE DIAGNOSTICS

The concept of a *knowledge space* is the central and most important concept in Doignon and Falmagne's theory. This is justified by the fact that several formally equivalent concepts are available (for instance the basis of the competence space and the corresponding surmise system) that are used for various purposes. Therefore, we now focus on the integration of the knowledge space concept into the competence-performance approach. In this section, we introduce the concept of a *union-stable diagnostic* and legitimize this concept by showing some of its properties that are of interest for practical application; furthermore we investigate the presuppositions for a goal-directed construction of a union-stable diagnostic.

Extending the knowledge space concept

Let $(\mathcal{E}, \mathcal{K}, \mathcal{A}, \mathcal{P}, k, p)$ be a diagnostic with a competence space $(\mathcal{K}, \cup)$. Common mathematical methods suggest postulating that the representation function p should be *union-preserving*, that is, p should map the union of some competence states to the union of the images of these states. An easily verified consequence is that the set $\mathcal{P}$ of performance states, as the image of $\mathcal{K}$ under the union-preserving function p, is also union-stable, with $p(\mathcal{E}) = \mathcal{A}$ being an element of $\mathcal{P}$; furthermore, it can be shown that $\emptyset = p(\emptyset)$ is in $\mathcal{P}$. Thus $(\mathcal{A}, \mathcal{P})$ is a *performance space* $(\mathcal{P}, \cup)$. This leads to the following definition.

DEFINITION 5.3 Let $(\mathcal{E}, \mathcal{K}, \mathcal{A}, \mathcal{P}, k, p)$ be a diagnostic, $(\mathcal{K}, \cup)$ a competence space, $(\mathcal{P}, \cup)$ a performance space, and $p : \mathcal{K} \longrightarrow \mathcal{P}$ a union-preserving representation function, that is:

$$p(\kappa \cup \lambda) = p(\kappa) \cup p(\lambda), \qquad \textit{for all } \kappa, \lambda \in \mathcal{K}.$$

Then the representation $(\mathcal{P}, \cup)$ of $(\mathcal{K}, \cup)$ is called a *union-preserving representation* of $(\mathcal{K}, \cup)$. The diagnostic $(\mathcal{E}, \mathcal{K}, \mathcal{A}, \mathcal{P}, k, p)$ is called a *union-stable diagnostic*.

REMARK 5.2 If the competence space $(\mathcal{K}, \cup)$ and the performance space $(\mathcal{P}, \cup)$ in a union-stable diagnostic $(\mathcal{E}, \mathcal{K}, \mathcal{A}, \mathcal{P}, k, p)$ are conceived as partial orderings $(\mathcal{K}, \subseteq)$ and $(\mathcal{P}, \subseteq)$ with the usual subset relation $\subseteq$ as the partial order, then the union-preserving representation function p is especially order-preserving. This fact is needed below.

The concept of a *union-stable diagnostic* reproduces the favorable properties of Doignon and Falgmagne's concept of a knowledge space on the competence as well as on the performance level; moreover, the competence space and the performance space in a union-stable diagnostic are structurally closely related by the union-preserving representation function. Thus, this concept seems an adequate extension of the concept of a knowledge space. Several properties of union-stable diagnostics are presented in the last subsection of this section. First, however, we will turn to a problem that is crucial for union-stable diagnostics: Which conditions can ensure that the representation function is union-preserving?

Clearly, because the representation function is completely determined by the interpretation function, these conditions will depend on the specific type of the problem set. We formulate the problem at hand as a representation problem:

> **Representation problem for competence spaces:**
>
> Given a competence space $(\mathcal{K}, \cup)$.
> Which conditions must a set $\mathcal{A}$ of problems or its interpretation $k(\mathcal{A}) := \{k_x \in \wp(\mathcal{K}) \mid x \in \mathcal{A}\}$ in $\mathcal{K}$ satisfy, so that the induced representation function $p : \mathcal{K} \longrightarrow \wp(\mathcal{A})$ generates a union-preserving representation $(\mathcal{P}, \cup)$ of $(\mathcal{K}, \cup)$, with $\mathcal{P} := p(\mathcal{K})$; in other words: that $(\mathcal{E}, \mathcal{K}, \mathcal{A}, \mathcal{P}, k, p)$ is a union-stable diagnostic?

In the following, we focus our attention on explicating conditions for the constructability of union-stable diagnostics. As is seen, these conditions can essentially be described as certain formal properties of problems.

Conditions of union-preserving representations

The following proposition reveals necessary and sufficient conditions for problems required for the construction of a union-stable diagnostic. (A proof can be found in Korossy, 1993.)

PROPOSITION 5.2 Let $(\mathcal{E}, \mathcal{K}, \mathcal{A}, \mathcal{P}, k, p)$ be a diagnostic, $(\mathcal{K}, \cup)$ a competence space. The representation function $p : \mathcal{K} \longrightarrow \mathcal{P}$ is union-stable, hence $(\mathcal{P}, \cup)$ a union-preserving representation of $(\mathcal{K}, \cup)$ and $(\mathcal{E}, \mathcal{K}, \mathcal{A}, \mathcal{P}, k, p)$ a union-stable diagnostic, if and only if for each problem $x \in \mathcal{A}$ and all κ, $\lambda \in \mathcal{K}$

(i) $\kappa \in k_x \wedge \kappa \subseteq \lambda \Longrightarrow \lambda \in k_x$,

(ii) $\kappa \cup \lambda \in k_x \Longrightarrow \kappa \in k_x \vee \lambda \in k_x$.

REMARK 5.3 Condition (i) in Proposition 5.2 means that k_x is an *order filter*. With the denotation

$$\uparrow Q := \{y \in P \mid \exists x \in Q(y \geq x)\}$$

for a partially ordered set $(P, \leq)$ and an arbitrary subset Q of P, and the denotation $\hat{k}_x :=$ Min k_x for the minimal elements in k_x, we could express Condition (i) in our concrete case through $k_x = \uparrow \hat{k}_x$.

Proposition 5.2 describes a very important fact: Conditions for a union-preserving representation of a competence space are *"local conditions"*, that means, conditions that each single problem must satisfy. In order to obtain a union-preserving representation function, no global property of the set of problems as a whole is required. This result is of great importance for the construction of a union-stable diagnostic:

> A union-stable diagnostic on a given competence space can be constructed step by step: If $(\mathcal{E}, \mathcal{K}, \mathcal{A}, \mathcal{P}, k, p)$ is a union-stable diagnostic and the set $\mathcal{A}$ of problems is extended to $\mathcal{A} \cup \{x\}$ by a problem x that satisfies the conditions of Proposition 5.2, then the resulting diagnostic is union-stable as well.

Union-stable problems

In the following, we analyze in more detail those problems that assure that the induced representation function is union-preserving. Let us first introduce a denotation for these types of problems.

DEFINITION 5.4 Let $(\mathcal{K}, \cup)$ be a competence space. A problem x interpreted in $\mathcal{K}$ through k_x is called *union-stable*, when for all $\kappa, \lambda \in \mathcal{K}$ conditions (i) and (ii) from Proposition 5.2 are satisfied.

According to Proposition 5.2 and Definition 5.4, the interpretation k_x of a *union-stable problem* x in a competence space $(\mathcal{K}, \cup)$ is characterized by the following two statements:

(1) If x is solvable in a certain competence state $\kappa \in \mathcal{K}$, then x is also solvable in each competence state $\lambda \in \mathcal{K}$ "above" κ (i.e. with $\kappa \subseteq \lambda$).
(2) If κ, λ are competence states in $\mathcal{K}$, and if the problem x is solvable in the competence state $\kappa \cup \lambda \in \mathcal{K}$, then the problem x is even solvable in *at least one* of these states κ *or* λ.

The characterization in (1) corresponds directly to our idea of a hierarchy of competence and performance states within a knowledge domain or a learning process that improves step by step. The characterization in (2) seems compatible with the idea of a union-preserving representation of the competence space. This idea includes that to each union of competence states should correspond the union of the assigned performance states. Thus, in any learning process that arrives at the union of the competence states, no problem should be solvable that was not solvable in at least one of the previous competence states.

Some criteria for union-stable problems

Because the central task in constructing a union-stable representation for a given competence space is the construction or selection of union-stable problems, in this subsection several specific characterizations of these types of problems for practical use will be explicated. Essentially, we present two theorems containing formal criteria for the interpretation k_x of a union-stable problem in a given competence space.[3] (For a competence structure $(\mathcal{E}, \mathcal{K})$ we use the short denotation $\mathcal{K}_\varepsilon$ for the family of all states in $\mathcal{K}$ containing an elementary competence $\varepsilon \in \mathcal{E}$.)

PROPOSITION 5.3 Let $(\mathcal{K}, \cup)$ be a competence space with the set $\mathcal{E}$ of elementary competencies. A problem x with interpretation $k_x \subseteq \mathcal{K}$ is a union-stable problem if and only if there exists a non-empty subset $\varphi \subseteq \mathcal{E}$ of elementary competencies, so that

$$k_x = \{\nu \in \mathcal{K} \mid \nu \cap \varphi \neq \emptyset\} = \bigcup \{\mathcal{K}_\delta \mid \delta \in \varphi\}.$$

Expressed somewhat differently, the statement for a union-stable problem x requires the existence of a non-empty subset $\varphi \subseteq \mathcal{E}$ of elementary competencies

[3] The proofs can be found in Korossy (1993).

so that for all $\nu \in \mathcal{K}$: $\nu \in k_x \Leftrightarrow \exists \varepsilon \in \varphi(\varepsilon \in \nu)$; thus a competence state belongs to k_x if and only if it contains at least one elementary competence of a set $\varphi \subseteq \mathcal{E}$ specific for x.

The following theorem provides a more convenient characterization for the interpretation k_x of a union-stable problem using the notion of an *order filter* (see Remark 5.3) and states from the basis or the surmise system of the competence space:

PROPOSITION 5.4 Let $(\mathcal{K}, \cup)$ be a competence space with the set $\mathcal{E}$ of elementary competencies, the basis $\mathcal{B}(\mathcal{K})$ and the surmise function σ.
A problem x with interpretation $k_x \subseteq \mathcal{K}$ is a union-stable problem if and only if there exists a non-empty subset $\varphi \subseteq \mathcal{E}$ of elementary competencies, such that for k_x one of the following equivalent conditions is satisfied:

$$\text{(1)}\quad k_x = \uparrow\{\beta \in \mathcal{B}(\mathcal{K}) \mid \beta \cap \varphi \neq \emptyset\} = \uparrow \mathrm{Min}\ \{\beta \in \mathcal{B}(\mathcal{K}) \mid \beta \cap \varphi \neq \emptyset\};$$
$$\text{(2)}\quad k_x = \uparrow\left(\bigcup\{\sigma(\delta) \mid \delta \in \varphi\}\right) = \uparrow \mathrm{Min}\ \left(\bigcup\{\sigma(\delta) \mid \delta \in \varphi\}\right).$$

Condition (1) of Proposition 5.4 means that the interpretation k_x of a union-stable problem x can be described as the order filter generated by those states from the basis $\mathcal{B}(\mathcal{K})$ that meet a problem-specific set φ of elementary competencies; condition (2) describes the union-stable problem x via the order filter generated by all clauses that are contained in at least one of the families $\sigma(\delta)$ for $\delta \in \varphi$. Obviously, in each of the two conditions the generating set for the order filter can be restricted to its minimal elements as is indicated by the second expression in each condition.

The various forms for the interpretation of a union-stable problem in a given competence space suggest the following notions.

DEFINITION 5.5 Let $(\mathcal{K}, \cup)$ be a competence space with the set $\mathcal{E}$ of elementary competencies; let x be a union-stable problem interpreted by k_x in $\mathcal{K}$.

(1) A non-empty subset $\varphi \subseteq \mathcal{E}$ of elementary competencies so that k_x can be described according to Proposition 5.4 is called a *generating set* for k_x.
(2) If φ is a generating set for k_x, then we call
$\tilde{k}_x := \{\beta \in \mathcal{B}(\mathcal{K}) \mid \beta \cap \varphi \neq \emptyset\}$ the *basis interpretation* of x,
$k_x^\sigma := \bigcup\{\sigma(\delta) \mid \delta \in \varphi\}$ the *surmise interpretation* of x.
(3) The unique subset $\hat{k}_x := \mathrm{Min}\ k_x$ is called the *minimal interpretation* of x (or k_x).

Let us state here that for the minimal, the basis, and the surmise interpretation of a union-stable problem x interpreted in $\mathcal{K}$ by k_x the following holds:

$$\hat{k}_x \subseteq k_x^\sigma \subseteq \tilde{k}_x \subseteq k_x, \quad \text{thus } \hat{k}_x = \mathrm{Min}\ k_x^\sigma = \mathrm{Min}\ \tilde{k}_x. \tag{2}$$

The defined notions of the *surmise interpretation* and the *basis interpretation* of a problem can apply as criteria for the union-stability of problems relative to a competence space $\mathcal{K}$: Under the precondition that the interpretation k_x of a problem x is accepted as being an order filter in $\mathcal{K}$, the problem x is union-stable in $\mathcal{K}$ if and only if x can be interpreted in $\mathcal{K}$ by one (and then every) of these two concepts. In contrast to that, the concept of the *minimal interpretation* does not uniquely characterize union-stable problems for a given competence space, but only problems necessary for an *order-preserving* representation function (Korossy, 1993). Nevertheless, the minimal interpretation of problems is very useful for an economical establishment of the representing performance space as is seen in the next subsection.

Properties of union-stable diagnostics

Let $(\mathcal{K}, \cup)$ be a competence space and $\mathcal{A}$ a problem set interpreted in $\mathcal{K}$ through a function p. In the preceding subsections, we described the conditions and some criteria which the problems in $\mathcal{A}$ must satisfy in order to generate a union-stable diagnostic $(\mathcal{E}, \mathcal{K}, \mathcal{A}, \mathcal{P}, k, p)$. In this subsection, some valuable properties of union-stable diagnostics are analyzed. We show for a union-stable diagnostic $(\mathcal{E}, \mathcal{K}, \mathcal{A}, \mathcal{P}, k, p)$

(a) how the representation function p can be computed using the basis of the competence space;
(b) how the basis of the performance space is constructed from the basis of the competence space;
(c) how the surmise function for the performance space is obtained from the surmise function of the competence space.

Let us begin with a readily apparent property of union-stable diagnostics that answers (a).

PROPOSITION 5.5 Let $(\mathcal{E}, \mathcal{K}, \mathcal{A}, \mathcal{P}, k, p)$ be a union-stable diagnostic and $\mathcal{B}(\mathcal{K})$ the basis of the competence space $\mathcal{K}$. If $\kappa \in \mathcal{K}$ is a competence state represented as $\kappa = \bigcup \mathcal{B}_\kappa(\mathcal{K})$ with an appropriate subset $\mathcal{B}_\kappa(\mathcal{K}) \subseteq \mathcal{B}(\mathcal{K})$ of the basis of $\mathcal{K}$, then the corresponding performance state $p(\kappa)$ is obtained as

$$p(\kappa) = \bigcup \{p(\beta) \mid \beta \in \mathcal{B}_\kappa(\mathcal{K})\} .$$

Proposition 5.5 states for a union-stable diagnostic its most important property. It allows the definition and computation of the representation function to be restricted to the basis of the competence space.

Proposition 5.5 suggests further considerations. Because each performance state $Z \in \mathcal{P}$ is the image of some competence state κ (see Definition 5.2 (2)),

Proposition 5.5 includes that each performance state can be written as a union of elements of $p(\mathcal{B}(\mathcal{K}))$; thus, $p(\mathcal{B}(\mathcal{K}))$ must contain the basis $\mathcal{B}(\mathcal{P})$ of the performance space, that is $\mathcal{B}(\mathcal{P}) \subseteq p(\mathcal{B}(\mathcal{K}))$. On the other hand, according to Remark 5.1, $\mathcal{B}(\mathcal{P})$ is obtained through $\mathcal{B}(\mathcal{P}) = \bigcup\{\hat{\mathcal{P}}_x \mid x \in \mathcal{A}\}$, where for all $x \in \mathcal{A}$ the family $\mathcal{P}_x$ includes all performance states containing x and $\hat{\mathcal{P}}_x :=$ Min $\mathcal{P}_x$ is the set of the minimal elements in $\mathcal{P}_x$. The question then is, how to select the families $\hat{\mathcal{P}}_x$ (which are subsets of $p(\mathcal{B}(\mathcal{K}))$) from $p(\mathcal{B}(\mathcal{K}))$.

According to Proposition 5.1, in each diagnostic $(\mathcal{E}, \mathcal{K}, \mathcal{A}, \mathcal{P}, k, p)$ the set $\mathcal{P}_x$ $(x \in \mathcal{A})$ can be constructed as the image $p(k_x)$ of the interpretation of x in $\mathcal{K}$ under p, that is $\mathcal{P}_x = p(k_x)$. If additionally $(\mathcal{E}, \mathcal{K}, \mathcal{A}, \mathcal{P}, k, p)$ is *union-stable*, hence the representation function especially *order-preserving* according to Remark 5.2, then it is easily proved for each x in $\mathcal{K}$ that $\hat{\mathcal{P}}_x \equiv \text{Min } p(k_x) \subseteq p(\text{Min } k_x) \equiv p(\hat{k}_x)$. Together with the subset relations (2) stated above, $\hat{k}_x \subseteq k_x^\sigma \subseteq \tilde{k}_x \subseteq k_x$, we obtain

$$\text{Min } p(k_x) \subseteq p(\hat{k}_x) \subseteq p(k_x^\sigma) \subseteq p(\tilde{k}_x) \subseteq p(k_x)\,,$$

and, as an immediate consequence, the important equalities

$$\hat{\mathcal{P}}_x \equiv \text{Min } p(k_x) = \text{Min } p(\hat{k}_x) = \text{Min } p(k_x^\sigma) = \text{Min } p(\tilde{k}_x)\,.$$

This result is of general interest for the answer to (b) as well as to (c). We summarize the preceding considerations by the following proposition.

PROPOSITION 5.6 A union-stable diagnostic $(\mathcal{E}, \mathcal{K}, \mathcal{A}, \mathcal{P}, k, p)$ includes the following properties:

(1) The basis $\mathcal{B}(\mathcal{P})$ of the performance space, which is obtained through $\mathcal{B}(\mathcal{P}) = \bigcup\{\hat{\mathcal{P}}_x \mid x \in \mathcal{A}\}$, is a subset of $p(\mathcal{B}(\mathcal{K}))$, that is $\mathcal{B}(\mathcal{P}) \subseteq p(\mathcal{B}(\mathcal{K}))$. The families $\hat{\mathcal{P}}_x$, which constitute the basis $\mathcal{B}(\mathcal{P})$ of the performance space, are obtained by taking the minimal elements from the image of the basis interpretation $\tilde{k}_x$ of x, that is $\hat{\mathcal{P}}_x = \text{Min } p(\tilde{k}_x)$.

(2) The surmise function $s : \mathcal{A} \longrightarrow \wp(\wp(\mathcal{A}))$ for the performance space is obtained from the surmise function σ for the competence space by collecting for each $x \in \mathcal{A}$ the minimal elements from the image of the surmise interpretation k_x^σ of x in the set $s(x)$, that is, by the equation $s(x) = \text{Min } p(k_x^\sigma)$ for each $x \in \mathcal{A}$.

(3) For the basis interpretation and the surmise interpretation used in (1) and (2), and for the minimal interpretation of an $x \in \mathcal{A}$, the following equalities hold:

$$\hat{\mathcal{P}}_x = \text{Min } p(\hat{k}_x) = \text{Min } p(k_x^\sigma) = \text{Min } p(\tilde{k}_x) = s(x)\,.$$

The consequences of these findings are of considerable practical use when a union-stable diagnostic has to be established:

Knowing the basis or the surmise function of the competence space and the interpretation of the union-stable problems on the basis of the competence space is sufficient for determining the representation function, and for constructing the basis and the surmise-function for the performance space, and thus the performance space itself.

EMPIRICAL APPLICATION

In order to test the applicability and utility of the theoretical framework introduced above, several empirical investigations were conducted, for example, within the field of elementary geometry (Korossy, 1993, 1996). In this section, we will report on a knowledge modeling in a specified field of elementary arithmetic that is currently the object of a series of validational studies, and that seems appropriate for illustrating the principle ideas of the modeling approach introduced here.

First, we briefly characterize the knowledge domain selected for the modeling studies. Next we describe the establishment of the competence modeling and the construction of the representing performance model. Then, we compare a small sample of observed solution patterns on a set of domain-specific problems with the family of hypothetically expected performance states. The disparity or correspondence between the solution patterns and the hypothesized performance states gives an indication regarding the empirical validity of the knowledge modeling.

Analysis of the knowledge domain

As a specific knowledge domain the area "divisibility in the set of natural numbers" was selected. This part of the elementary number theory is dealt with in nearly every textbook on elementary arithmetic, especially because of its importance in connection with or as a prerequisite for teaching fractions. In general, the following basic topics of instruction in the selected knowledge domain are treated:

- divisors and multiples
- propositions and rules of divisibility
- prime numbers and composites
- prime factorization of natural numbers (PF)
- greatest common divisor (GCD)
- least common multiple (LCM)

(see e.g. Setek, 1992; Schmid, 1994; Falstein, 1986; Meserve & Sobel, 1989).

Our knowledge modeling for the area "divisibility in the set of natural numbers" is based on the curricular structure that is followed in most of the common textbooks introducing elementary arithmetics. Table 1 describes several central instructional topics of this curriculum in the form of computational methods. (Each method is designated by a certain abbreviation for later use.) The methods listed might easily be formulated as teaching/learning goals of the curriculum as well. They define the qualitative, essentally procedural knowledge elements (or "skills") of the selected field with respect to the intended application.[4]

The next task is an analysis of the knowledge domain with respect to diagnostically relevant prerequisite relations leading lastly to a surmise function on the selected instructional topics. Table 1 provides some indication for that task; however, the defined topics, per se, do not determine the required order structure completely. The main reason for this is that the diagnostically relevant dependencies between the topics are of different categorial types. At least three different types of relations should be distinguished in the selected knowledge domain:

1. logical/mathematical relations
2. relations concerning differences in complexity/difficulty
3. relations based on instructional experience.

The following examples may demonstrate what types and to what extent these various types of relations are significant for a diagnostical prerequisite relation:

EXAMPLE 1a: A logical/mathematical prerequisite for the computation of the greatest common divisor of two natural numbers by the *method using prime factorizations* (Table 1: G_P) is, naturally, the computation of the prime factorization of a natural number (Table 1: P).

EXAMPLE 1b: The computation of the set of all common divisors of two natural numbers using method 1 (Table 1: C_D) presupposes the computation of the sets of all divisors for the two numbers. This, however, can be accomplished by two alternative methods (Table 1: D_P *or* D_F). Thus, we have two alternative logical/mathematical prerequisites.

EXAMPLE 2: It seems more difficult to determine the set of all divisors for the natural number 230 than for 77. This is justified by looking at the two available methods D_F and D_P for computing the set of divisors of a natural number: In each of the two methods, more computational steps are necessary for determining the set of all divisors for 230 than are necessary for 77. This difference in complexity/difficulty is also reflected by the fact that the set of all divisors of 230 contains eight elements, whereas 77 has only four divisors.

EXAMPLE 3: In most courses in the area of divisibility, the concept and methods for the computation of the least common multiple of natural numbers are

[4] In our special case, for example, the target group-directed and context-oriented knowledge modeling suggested leaving out one elementary method for computing the least common multiple $\mathrm{LCM}(a, b)$ of two natural numbers a, b:
(i) List the set of multiples for each of the numbers a and b.
(ii) Select the least of the multiples appearing in both sets.
This method is omitted in the modeling presented here because it had not been extensively practiced in the target group of our study.

TABLE 1
Teaching-Learning Objectives of the Knowledge Domain "Divisibility in the Set of Natural Numbers"

Computing the prime factorization of a natural number a			
P	*Method*:		Factorize a (stepwise) until all factors are prime. The result is the prime factorization PF(a) of a: $a = \prod_p p^{\alpha(p)}$, where p ranges over all prime numbers, and $\alpha(p)$ is the multiplicity of factor p.
Computing the set of all divisors of a natural number a			
D_F	*Method 1*:	(i)	Find all decompositions $a = u \cdot v$, with $u,\ v \in N$.
		(ii)	For each product $a = u \cdot v:\ u,\ v \in D_a$.
D_P	*Method 2*:	(i)	Compute PF(a).
		(ii)	Collect in D_a 1, all prime factors and all possible products made up of prime factors of a.
Computing the set of common divisors of two natural numbers a, b			
C_D	*Method 1*:	(i)	Compute the sets D_a, D_b of divisors of a, b.
		(ii)	Compute the intersection $D_a \cap D_b$.
C_G	*Method 2*:	(i)	Compute GCD(a, b) of the two numbers a, b.
		(ii)	Compute the set of divisors $D_{\mathrm{GCD}(a,b)}$.
Computing the greatest common divisor of two natural numbers a, b			
G_D	*Method 1*:	(i)	Compute the intersection set $D_a \cap D_b$.
		(ii)	Select the greatest element $\max(D_a \cap D_b)$.
G_P	*Method 2*:	(i)	Compute PF(a) and PF(b): $a = \prod_p p^{\alpha(p)}$ and $b = \prod_p p^{\beta(p)}$.
		(ii)	Compute $\mathrm{GCD}(a,b) = \prod_p p^{\min(\alpha(p),\beta(p))}$.
Computing the least common multiple of two natural numbers a, b			
L_P	*Method 1*:	(i)	Compute PF(a) and PF(b): $a = \prod_p p^{\alpha(p)}$ and $b = \prod_p p^{\beta(p)}$.
		(ii)	Compute $\mathrm{LCM}(a,b) = \prod_p p^{\max(\alpha(p),\beta(p))}$.
L_G	*Method 2*:	(i)	Compute $\mathrm{GCD}(a,b)$.
		(ii)	Compute $\mathrm{LCM}(a,b) = ab/\mathrm{GCD}(a,b)$.

D_a	the *set of all divisors* of $a \in N$
PF(a)	the *prime factorization* of $a \in N$
GCD(a, b)	the *greatest common divisor* of $a, b \in N$
LCM(a, b)	the *least common multiple* of $a, b \in N$

introduced only after computational methods for determining the greatest common divisor have been practiced sufficiently. This sequence is legitimized by the instructional experience that for most pupils the prime factor method for computing the least common multiple has proven to be more difficult to apply than the prime factor method for determining the greatest common divisor.

For our modeling of the chosen knowledge domain, the majority of the implemented diagnostically relevant relations are founded in logical/mathematical dependencies. However, in order to arrive at a sufficiently precise surmise function, the following standardization concerning the complexity/difficulty was imposed on the set of topics:

1. All teaching/learning goals (with two exceptions) apply to natural numbers with structurally equivalent prime factorizations of the form $a = p_1 \cdot p_2 \cdot p_3$ with $(p_1, p_2, p_3) \in \{2, 3, 5, 7\}^2 \times \{11, 13, 17, 19, 23, 29, 31, 37, 41, 43\}$.
2. Only for each of the two topics D_P and D_F, additionally a lower grade of complexity/difficulty is included, denoted by D_P^- resp. D_F^-, and characterized by the application to natural numbers with prime factor structure $a = p_1 \cdot p_2$ with $(p_1, p_2) \in \{2, 3, 5, 7\} \times \{11, 13, 17, 19, 23, 29, 31, 37, 41, 43\}$.

Altogether, only two single diagnostical relations in our knowledge modeling explicitly concern different grades in complexity/difficulty within one topic, whereas all other diagnostical relations concern different topics. Aside from the logically/mathematically based dependencies between topics, we have implemented one relation that is justified by didactical experience in the sense of Example 3. Table 2 shows the hypothesized diagnostically significant relations between successive topics of Table 1. It seems reasonable to presuppose that the relations shown in Table 2 can be transitively extended. Note that all the explicated relations bear a hypothetical character up until this point and will be the subject of empirical validation.

In the following two subsections, the formal establishment of the competence model and the performance representation is described.

Modeling the competence space

Formally, a computational method can be identified with the ability or the competency of performing this method. Therefore, in our competence modeling the teaching/learning objectives defined in Table 1, and, additionally, the lower grades of difficulty D_P^- and D_F^- of D_P and D_F, were conceived as *elementary competencies*. That is, we have constituted the family

$$\mathcal{E} = \{P, D_P^-, D_F^-, D_P, D_F, C_D, C_G, G_D, G_P, L_P, L_G\}$$

TABLE 2
Diagnostically Significant Relations
Between the Teaching/Learning Objectives and Their Justification.

Type	*Relations*
1. logical/mathematical	$D_P \rightarrow P$
	$C_D \rightarrow D_P \vee D_F$
	$C_G \rightarrow G_P$; $C_G \rightarrow D_{\bar{P}} \vee D_{\bar{F}}$
	$G_P \rightarrow P$
	$G_D \rightarrow C_D$
	$L_G \rightarrow G_D \vee G_P$
2. complexity/difficulty	$D_P \rightarrow D_{\bar{P}}$; $D_F \rightarrow D_{\bar{F}}$
3. instructional experience	$L_P \rightarrow G_P$

$\rightarrow$ symbolizes the surmise relation. Only relations between successive topics are listed. For example, $L_G \rightarrow G_D \vee G_P$ means: If a person is capable of performing method L_G, then the person is surmised to be capable of performing method G_D *or* method G_P as well.

TABLE 3
Surmise Function $\sigma : \mathcal{E} \longrightarrow \wp(\wp(\mathcal{E}))$ for the Competence Space

ε	$\sigma(\varepsilon)$	ε	$\sigma(\varepsilon)$
P	$\{P\}$	C_D	$\{PD_{\bar{P}}D_PC_D, D_{\bar{F}}D_FC_D\}$
$D_{\bar{P}}$	$\{D_{\bar{P}}\}$	C_G	$\{PD_{\bar{P}}C_GG_P, PD_{\bar{F}}C_GG_P\}$
$D_{\bar{F}}$	$\{D_{\bar{F}}\}$	G_D	$\{PD_{\bar{P}}D_PC_DG_D, D_{\bar{F}}D_FC_DG_D\}$
D_P	$\{PD_{\bar{P}}D_P\}$	G_P	$\{PG_P\}$
D_F	$\{D_{\bar{F}}D_F\}$	L_P	$\{PG_PL_P\}$
		L_G	$\{PG_PL_G, PD_{\bar{P}}D_PC_DG_DL_G, D_{\bar{F}}D_FC_DG_DL_G\}$

of 11 elementary competencies.

The most natural and immediate approach to the modeling of the competence states, in the present case, is through the surmise function. Suggested by Table 2, the *surmise function* $\sigma : \mathcal{E} \longrightarrow \wp(\wp(\mathcal{E}))$ can be established; it is explicitly defined by Table 3[5]. One easily checks that for the clauses of σ the three conditions (1), (2), (3) for a surmise system given in the first section are satisfied.

According to Doignon and Falmagne's theorem (see the first section), the

[5] For a shorter notation, we write the clauses as sequences of symbols instead of as sets. For instance, for the family

$$\Big\{ \{P, G_P, L_G\}, \{P, D_{\bar{P}}, D_P, C_D, G_D, L_G\}, \{D_{\bar{F}}, D_F, C_D, G_D, L_G\} \Big\}$$

of clauses for L_G, we use the short notation

$$\Big\{ PG_PL_G, PD_{\bar{P}}D_PC_DG_DL_G, D_{\bar{F}}D_FC_DG_DL_G \Big\}.$$

surmise system $(\mathcal{E}, \sigma)$ determines uniquely a *competence space* $(\mathcal{K}, \cup)$. This one-to-one correspondence is specified through the equation $\sigma(\varepsilon) = \hat{\mathcal{K}}_\varepsilon$ for all $\varepsilon \in \mathcal{E}$, where $\mathcal{K}_\varepsilon$ denotes the subset of all states of $\mathcal{K}$ containing an $\varepsilon \in \mathcal{E}$, and $\hat{\mathcal{K}}_\varepsilon := \text{Min } \mathcal{K}_\varepsilon$ denotes the minimal elements in each $\mathcal{K}_\varepsilon$. The family of all clauses occurring in Table 3 determines the basis $\mathcal{B}(\mathcal{K})$ of the competence space $(\mathcal{K}, \cup)$ via $\mathcal{B}(\mathcal{K}) = \bigcup\{\sigma(\varepsilon) \mid \varepsilon \in \mathcal{E}\} \equiv \bigcup\{\hat{\mathcal{K}}_\varepsilon \mid \varepsilon \in \mathcal{E}\}$, and the family $\mathcal{K}$ of competence states is given by the closure of $\mathcal{B}(\mathcal{K})$ under union.

The *basis* $\mathcal{B}(\mathcal{K})$ of $\mathcal{K}$ consists of 16 competence states. The entire competence space $(\mathcal{K}, \cup)$ contains 184 competence states; these are 8.98% of the power set $\wp(\mathcal{E})$ with 2048 elements.

Constructing a performance representation

As a first step in establishing a representing performance structure for $(\mathcal{K}, \cup)$, a set $\mathcal{A} := \{a, b, c, d, e, f\}$ containing six problems of the area "divisibility in the set of natural numbers" was constructed. Table 4 shows the six problems of set $\mathcal{A}$ together with their solution ways.

It is immediately apparent that each solution way of a problem refers directly to a particular method listed in Table 1 (as is indicated by the respective abbreviation; problem a refers to the methods D_F^- and D_P^- that are the lower grades of difficulty of the methods D_F and D_P). Furthermore, in accordance with the definition of $\mathcal{E}$, each solution way corresponds to a particular elementary competency of $\mathcal{E}$. This correspondence between solution ways and elementary competencies of $\mathcal{E}$ suggests an *interpretation function* $k : \mathcal{A} \longrightarrow \wp(\mathcal{K})$ for the set of problems within the family $\mathcal{K}$ of competence states by taking the set of solution ways for each problem $x \in \mathcal{A}$ as a *generating set* (see Definition 5.5) for the interpretation k_x of this problem x. This interpretation presupposes, for example, for problem $e \in \mathcal{A}$, which can be solved using method G_D *or* method G_P, that e should be solvable in each competence state containing at least the elementary competency G_D *or* the elementary competency G_P; that is, $\{G_D, G_P\}$ is taken as a generating set for the interpretation k_e of e in $\mathcal{K}$. Generating sets for the other problems of set $\mathcal{A}$ can be seen from Table 4.

Table 5 shows for each problem $x \in \mathcal{A}$ the *surmise interpretation* k_x^σ, and for the purpose of illustration additionally the *basis interpretation* $\tilde{k}_x$ and the *minimal interpretation* $\hat{k}_x$ of the problem, whereby it is presupposed that $k_x = \uparrow\tilde{k}_x \equiv \uparrow k_x^\sigma \equiv \uparrow\hat{k}_x$. For example, the surmise interpretation k_e^σ of the problem $e \in \mathcal{A}$ with generating set $\{G_D, G_P\}$ is the set of all clauses occurring in $\sigma(G_D) \cup \sigma(G_P)$ (see Table 3); the basis interpretation $\tilde{k}_e$ of e is the set of all states from the basis $\mathcal{B}(\mathcal{K})$ that contain G_D or G_P.[6] According to Proposition 5.4, each

[6] Table 5 confirms that for each problem x holds $\hat{k}_x \subseteq k_x^\sigma \subseteq \tilde{k}_x$, which was stated in (2) in connection with Definition 5.5. In this case, however, the surmise interpretation coincides with the minimal interpretation.

TABLE 4

The Six Problems on Divisibility and Their Solution Ways

Problem a : Compute the set of all divisors of 77.			
D_F^-	*Method 1*:	(i)	$77 = 1 \cdot 77 = 7 \cdot 11$
		(ii)	$D_{77} = \{1, 7, 11, 77\}$.
D_P^-	*Method 2*:	(i)	$77 = 7 \cdot 11$
		(ii)	Divisors: 1, 7, 11, 77, thus $D_{77} = \{1, 7, 11, 77\}$.
Problem b : Compute the set of all divisors of 230.			
D_F	*Method 1*:	(i)	$230 = 1 \cdot 230 = 2 \cdot 115 = 5 \cdot 46 = 10 \cdot 23$
		(ii)	$D_{230} = \{1, 2, 5, 10, 23, 46, 115, 230\}$.
D_P	*Method 2*:	(i)	$230 = 2 \cdot 5 \cdot 23$
		(ii)	Divisors: 1, 2, 5, 23, 10, 46, 115, 230; thus $D_{230} = \{1, 2, 5, 10, 23, 46, 115, 230\}$.
Problem c : Compute the prime factorization of 273.			
P	*Method*:		$273 = 3 \cdot 91 = 3 \cdot 7 \cdot 13$
Problem d : Compute the set of common divisors of 172 and 258.			
C_D	*Method 1*:	(i)	$D_{172} = \{1, 2, 4, 43, 86, 172\}$ $D_{258} = \{1, 2, 3, 6, 43, 86, 129, 258\}$
		(ii)	$D_{172} \cap D_{258} = \{1, 2, 43, 86\}$.
C_G	*Method 2*:	(i)	$\mathrm{GCD}(172, 258) = 86$
		(ii)	$D_{172} \cap D_{258} = D_{86} = \{1, 2, 43, 86\}$.
Problem e : Compute the greatest common divisor of 275 and 385.			
G_D	*Method 1*:	(i)	$D_{275} \cap D_{385} = \{1, 5, 11, 55\}$
		(ii)	$\mathrm{GCD}(275, 385) = \max(D_{275} \cap D_{385}) = 55$.
G_P	*Method 2*:	(i)	$275 = 5 \cdot 5 \cdot 11$; $385 = 5 \cdot 7 \cdot 11$
		(ii)	$\mathrm{GCD}(275, 385) = 5 \cdot 11 = 55$.
Problem f : Compute the least common multiple of 275 and 385.			
L_P	*Method 1*:	(i)	$275 = 5 \cdot 5 \cdot 11$; $385 = 5 \cdot 7 \cdot 11$
		(ii)	$\mathrm{LCM}(275, 385) = 5 \cdot 5 \cdot 7 \cdot 11 = 1925$.
L_G	*Method 2*:	(i)	$\mathrm{GCD}(275, 385) = 55$
		(ii)	$\mathrm{LCM}(275, 385) = (275 \cdot 385)/55 = 1925$.

Denotations as in Table 1

TABLE 5
Minimal, Surmise, and Basis Interpretation
for the Problems, and the Representation Function $p : \mathcal{B}(\mathcal{K}) \longrightarrow \wp(\mathcal{A})$

$\beta \in \mathcal{B}(\mathcal{K})$	*Interpretation of the problems* a	b	c	d	e	f	$p(\beta)$
P			$\circledast$				c
D_P^-	$\circledast$						a
D_F^-	$\circledast$						a
$PD_P^-D_P$	$+$	$\circledast$	$+$				abc
$D_F^-D_F$	$+$	$\circledast$					ab
$PD_P^-D_PC_D$	$+$	$+$	$+$	$\circledast$			$abcd$
$D_F^-D_FC_D$	$+$	$+$		$\circledast$			abd
$PD_P^-C_GG_P$	$+$		$+$	$\circledast$	$+$		$acde$
$PD_F^-C_GG_P$	$+$		$+$	$\circledast$	$+$		$acde$
$PD_P^-D_PC_DG_D$	$+$	$+$	$+$	$+$	$\circledast$		$abcde$
$D_F^-D_FC_DG_D$	$+$	$+$		$+$	$\circledast$		$abde$
PG_P			$+$		$\circledast$		ce
PG_PL_P			$+$		$+$	$\circledast$	cef
PG_PL_G			$+$		$+$	$\circledast$	cef
$PD_P^-D_PC_DG_DL_G$	$+$	$+$	$+$	$+$	$+$	$\circledast$	$abcdef$
$D_F^-D_FC_DG_DL_G$	$+$	$+$		$+$	$+$	$\circledast$	$abdef$

$+$ basis interpretation $\tilde{k}_x$ of problem $x \in \mathcal{A}$
$\times$ surmise interpretation k_x^σ of $x \in \mathcal{A}$
$\odot$ minimal interpretation $\hat{k}_x$ of $x \in \mathcal{A}$.

problem $x \in \mathcal{A}$ is union-stable.

The interpretation function k uniquely determines the *representation function* $p : \mathcal{K} \longrightarrow \wp(\mathcal{A})$, which is union-preserving as a consequence of the union-stable problems (by Proposition 5.2 and Definition 5.4). By Proposition 5.5, p can be restricted to the basis $\mathcal{B}(\mathcal{K})$ of the competence space $(\mathcal{K}, \cup)$. This restricted representation function $p : \mathcal{B}(\mathcal{K}) \longrightarrow \wp(\mathcal{A})$ can be gathered from the last column of Table 5.

Because union-stable problems satisfy (i) and (ii) of Proposition 5.2 via Definition 5.4, the performance structure $(\mathcal{A}, \mathcal{P})$, with $\mathcal{P} := p(\mathcal{K})$, is a *performance space* $(\mathcal{P}, \cup)$. Two parallel ways, both described by Proposition 5.6, can be followed for the establishment of $\mathcal{P}$. One way uses the basis of $(\mathcal{P}, \cup)$, the other way uses the surmise function $s : \mathcal{A} \longrightarrow \wp(\wp(\mathcal{A}))$ of the performance space.

According to Proposition 5.6 (1), the *basis* $\mathcal{B}(\mathcal{P})$ of $(\mathcal{P}, \cup)$ can be found within the images of $\mathcal{B}(\mathcal{K})$ under the representation function p (see Table 5): For each $x \in \mathcal{A}$ select from the image set $p(\mathcal{B}(\mathcal{K}))$ the family $p(\tilde{k}_x) \subseteq p(\mathcal{B}(\mathcal{K}))$ and take the minimal states of $p(\tilde{k}_x)$. According to Proposition 5.6 (1), $\hat{\mathcal{P}}_x =$

Min $p(\tilde{k}_x)$. Now, the basis $\mathcal{B}(\mathcal{P})$ of the performance space is immediately obtained by $\mathcal{B}(\mathcal{P}) = \bigcup\{\hat{\mathcal{P}}_x \mid x \in \mathcal{A}\}$ from Table 5:

$$\mathcal{B}(\mathcal{P}) = \{a, ab, c, abd, acde, abde, ce, cef, abdef\}\,.$$

A comparable way leads to the establishment of the *surmise function* $s : \mathcal{A} \longrightarrow \wp(\wp(\mathcal{A}))$ for the performance space. In accordance with Proposition 5.6 (2), s is constructed from the surmise function σ of the competence space by putting for each $x \in \mathcal{A}$ the minimal elements from the image of the surmise interpretation k_x^σ of x (see Table 5) into the set $s(x)$. That procedure is described by the equation $s(x) = \text{Min}\ p(\tilde{k}_x^\sigma)$ for each $x \in \mathcal{A}$. Table 6 shows for each $x \in \mathcal{A}$ the families $p(\tilde{k}_x)$ (from Table 5) and $s(x) = \text{Min}\ p(\tilde{k}_x)$.

It should be remarked at this point that, generally, the most economical procedure for obtaining $\hat{\mathcal{P}}_x$ or $s(x)$ is at first to form the minima $\hat{k}_x \equiv \text{Min}\ (\tilde{k}_x) \equiv \text{Min}\ (k_x^\sigma)$ of the problem interpretation k_x, and only after that to compute $\hat{\mathcal{P}}_x \equiv \text{Min}\ p(\hat{k}_x) \equiv s(x)$. This procedure is justified by Proposition 5.6 (3). However, in our present case, the minimal interpretation $\hat{k}_x$ and the surmise interpretation $\tilde{k}_x^\sigma$ happen to coincide for each problem $x \in \mathcal{A}$; thus, the procedure demonstrated by Table 6 is the most economical one for generating the families $\hat{\mathcal{P}}_x$ or $s(x)$.

By the basis $\mathcal{B}(\mathcal{P})$ and, equivalently, by the surmise function s the *performance space* $(\mathcal{P}, \cup)$ is uniquely determined. The performance space $(\mathcal{P}, \cup)$ contains 20 performance states. These are 31.25% of all possible solution patterns in the power set $\wp(\mathcal{A})$ of $\mathcal{A}$.

The hypothesis for the data analysis was that the observed solution patterns on the set $\mathcal{A}$ of problems agree with the theoretically expected solution patterns, that means with the performance states of $\mathcal{P}$.

TABLE 6
Construction of the Families $s(x) = \text{Min}\ p(\tilde{k}_x)$
Using the Surmise Interpretations $\tilde{k}_x^\sigma$ of the Problems $x \in \mathcal{A}$

$x \in \mathcal{A}$	$p(\tilde{k}_x)$	$s(x) = \text{Min}\ p(\tilde{k}_x)$
a	$\{a, a\}$	$\{a\}$
b	$\{abc, ab\}$	$\{ab\}$
c	$\{c\}$	$\{c\}$
d	$\{abcd, abd, acde, acde\}$	$\{abd, acde\}$
e	$\{abcde, abde, ce\}$	$\{abde, ce\}$
f	$\{cef, cef, abcdef, abdef\}$	$\{cef, abdef\}$

TABLE 7
The Six Problems of the Set $\mathcal{A}'$ for Group R

Problem a' :	Compute the set of all divisors of 93.
Problem b' :	Compute the set of all divisors of 222.
Problem c' :	Compute the prime factorization of 266.
Problem d':	Compute the set of common divisors of 165 and 363.
Problem e' :	Compute the greatest common divisor of 153 and 357.
Problem f' :	Compute the least common multiple of 153 and 357.

Method

Problems

For each of two parallel test groups, designated as L and R, a set of six problems was constructed. The problems for group L are collected in the set $\mathcal{A} := \{a, b, c, d, e, f\}$ and are shown by Table 4 together with their possible solution ways. The six problems of the set $\mathcal{A}' = \{a', b', c', d', e', f'\}$ for group R are presented in Table 7.

The reader may verify that each problem $x' \in \mathcal{A}'$ is solvable by using the same methods as for the solution of problem $x \in \mathcal{A}$. Moreover, the prime factorizations of the included numbers are in accordance with the established standardization.[7] Thus, the problems of $\mathcal{A}'$ can be interpreted in $\mathcal{K}$ through an interpretation function $k' : \mathcal{A}' \longrightarrow \wp(\mathcal{K})$, where $k'(x') = k(x)$ for each pair (x, x') with $x \in \mathcal{A}$, $x' \in \mathcal{A}'$. Following the argumentation in connection with (2) of Proposition 5.1, the problems of $\mathcal{A}$ and $\mathcal{A}'$ are considered pairwise equivalent. Regarded formally, using the problem set $\mathcal{A}'$ we obtain another diagnostic $(\mathcal{E}, \mathcal{K}, \mathcal{A}', \mathcal{P}', k', p')$, which is structurally identical to $(\mathcal{E}, \mathcal{K}, \mathcal{A}, \mathcal{P}, k, p)$. A comparison of the two representations, however, was not the object of our empirical study; rather the two equivalent problem sets were constructed with respect to the practical demands of group testing (see above) and are conceived as two equivalent versions of the same problem type.

The complete test for each of the two groups consisted of the six problems concerning the knowledge domain "divisibility in the set of natural numbers", and, additionally, three problems from elementary geometry that were not included in the evaluation of this study.

[7] The only exception is problem d', where $363 = 3 \cdot 11 \cdot 11$, which was not crucial for the test results.

Subjects

The test was conducted in a class of a German high school. The 23 pupils (13 male, 10 female), between 11 and 12 years of age, had been taught elementary number theory for about 25 hours as part of regular mathematics lessons. This area is part of a second year German high school mathematics curriculum set down by the Department of Education (Ministerium für Kultus und Sport, 1994).

Procedure

The test was conducted during a regular mathematics lesson. It was carried out as a group test. In order to avoid "transfer effects", the class was divided into two parallel groups L (11 persons) and R (12 persons); pupils sitting side by side were assigned to different groups. Each group worked on one of two equivalent sets of eight problems. For each group, the eight problems were presented in the form of a two-sided test. The calculations and solution ways were to be filled in in the blank spaces provided. The problems could be worked on in any order. The time limit for the entire test was 40 minutes.

Results

Although the theoretical approach introduced in this paper presupposes only two-categorial solution behavior (correct/incorrect), the test procedure (paper and pencil) provided additional information on the applied solution approaches and solution ways. This information was registered and evaluated with regard to the possibility of validating the solution analyses of the problems. Nevertheless, for the evaluation of the results as *correct* or *incorrect*, a strict criterion was applied. For instance, a set of divisors was judged to be correct only if it contained *all divisors*. Table 8 shows for each person of group L and group R the observed solution way for each problem and the solution pattern for the entire set of problems.[8]

As can be seen from Table 8, for all 11 subjects of group L, the solution patterns are in accordance with expected performance states. In group R, in all two non-expected solution patterns out of the 12 solution patterns are observed. In the set of solution patterns of group L, seven different performance states occur. In group R, six different expected solution patterns, that is performance states, are found, whereas the two other patterns are not in accordance with the expectation.

[8] The results for group L and group R are separated in Table 8 in order to control the presupposed equivalence of the two problem sets, although this is not the object of the data analysis.

TABLE 8
Solution Patterns for Group L on the Problems
a, b, c, d, e, f and for Group R on the Problems a', b', c', d', e', f'

	Solution way observed in the test						*Solution*	*State?*
Subj.	a, a	b, b'	c, c'	d, d'	e, e'	f, f'	*pattern*	+/□
02 L	D_P^-	D_P	P	–	G_P	L_P	*abcef*	+
03 L	*	*	P	C_D	G_P	L_P	*abcdef*	+
05 L	D_P^-	D_P	P	C_D	G_P	–	*abcde*	+
08 L	*	–	P	–	–	–	*ac*	+
11 L	*	*	P	–	G_P	–	*abce*	+
13 L	*	–	P	–	G_P	L_P	*acef*	+
17 L	D_P^-	D_P	P	C_G	G_P	L_P	*abcdef*	+
20 L	D_P^-	D_P	P	–	–	–	*abc*	+
21 L	D_P^-	D_P	P	–	–	–	*abc*	+
22 L	*	–	P	–	–	–	*ac*	+
23 L	*	–	P	–	G_P	L_P	*acef*	+
01 R	D_P^-	–	P	C_G	G_P	L_P	$a'c'd'e'f'$	+
04 R	*	*	–	–	–	–	$a'b'$	+
06 R	D_P^-	–	P	–	–	–	$a'c'$	+
07 R	–	D_P	–	C_D	G_P	L_P	$b'd'e'f'$	□
09 R	D_P^-	D_P	P	–	–	–	$a'b'c'$	+
10 R	–	–	P	–	–	–	c'	+
12 R	D_P^-	D_P	P	–	G_P	–	$a'b'c'e'$	+
14 R	*	*	P	–	–	–	$a'b'c'$	+
15 R	D_F^-	D_F	P	–	–	L_P	$a'b'c'f'$	□
16 R	D_F^-	–	P	–	–	–	$a'c'$	+
18 R	*	–	P	–	–	–	$a'c'$	+
19 R	*	–	P	–	–	–	$a'c'$	+

– The problem is not solved or not solved correctly.
* The result is correct, but the solution way is not uniquely identifiable.
+ The solution pattern coincides with a performance state.
□ The solution pattern does not coincide with a performance state.

Discussion

The knowledge modeling study on *divisibility in the set of natural numbers* was presented primarily for the purpose of illustrating a practical application of the theoretical concepts of the competence-performance approach outlined in the preceding sections. The selected knowledge domain has been modeled in the form of a competence space and an empirical representation in the form of a performance space.

Within the knowledge domain in question there exist both curricular networks of teaching/learning objectives and a large pool of problems in text books on elementary arithmetic. The analysis of domain-specific teaching/learning objectives provided a family $\mathcal{E}$ of elementary competencies, and a surmise function σ that imposes a diagnostically relevant structure on the elementary competencies. By these two components a competence space $(\mathcal{K}, \cup)$ is defined. In order to obtain an empirical representation of $(\mathcal{K}, \cup)$, a small set of six appropriate problems was constructed (each problem in two equivalent versions) and interpreted in $\mathcal{K}$, so that a union-stable diagnostic for the knowledge domain resulted.

In principle, the reported analysis of an empirical data set can be regarded as a validation study for the established competence space (and the interpretation of the problems in $\mathcal{K}$). Certainly, because of the small sample of solution patterns, the results of the study are not very meaningful; nevertheless, the high congruence of the empirically observed and the theoretically expected solution patterns involves some evidence for the psychological validity of the underlying competence model (and task analyses).[9]

Should there be further interest in the modeled competence space, the validation process would have to be continued by establishing another performance representation and testing the performance states against the observed solution patterns. This, in fact, is currently being done in several validation studies using slightly different variants of the modeled competence space and other domain-specific problem sets. The results are reported elsewhere.

Whenever a competence space is accepted as being psychologically valid, then, in connection with a suitable performance representation, it can be used as a diagnostical framework for an economic qualitative competence diagnosis and goal-oriented adaptive learning/teaching processes. Moreover, because of the union-stable structures on the competence and the performance space and the union-preserving representation function, much of the practical work can be done in a convenient and economical way by making use of the equivalent concepts of the (competence/performance) space, the basis of the space, and the according surmise system.

ACKNOWLEDGEMENTS

The research reported in this paper was supported by Grant Lu 385/1 of the Deutsche Forschungsgemeinschaft to J. Lukas and D. Albert at the University of Heidelberg.

[9] At this point we do not discuss the problems concerning reasonable test models for a statistically unobjectionable evaluation of the empirical results.

REFERENCES

Albert, D., & Held, T. (1994). Establishing knowledge spaces by systematical problem construction. In D. Albert (Ed.), *Knowledge structures* (pp. 81–115). Berlin, Heidelberg: Springer.

Doignon, J.-P. (1994a). Probabilistic assessment of knowledge. In D. Albert (Ed.), *Knowledge structures* (pp. 1–57). Berlin: Springer.

Doignon, J.-P. (1994b). Knowledge spaces and skill assignments. In G. H. Fischer & D. Laming (Eds.), *Contributions to mathematical psychology, psychometrics, and methodology* (pp. 111–121). Berlin: Springer.

Doignon, J.-P., & Falmagne, J.-C. (1985). Spaces for the assessment of knowledge. *International Journal of Man-Machine Studies, 23*, 175–196.

Doignon, J.-P., & Falmagne, J.-C. (1998). *Knowledge spaces*. Berlin: Springer.

Dowling, C. E. (1993). Applying the basis of a knowledge space for controlling the questioning of an expert. *Journal of Mathematical Psychology, 37*, 21–48.

Dowling, C. E. (1994). Combinatorial structures for the representation of knowledge. In D. Albert (Ed.), *Knowledge structures* (pp. 59–79). Berlin: Springer.

Düntsch, I., & Gediga, G. (1995). Skills and knowledge structures. *British Journal of Mathematical and Statistical Psychology, 48*, 9–27.

Falmagne, J.-C., & Doignon, J.-P. (1988a). A class of stochastic procedures for the assessment of knowledge. *British Journal of Mathematical and Statistical Psychology, 41*, 1–23.

Falmagne, J.-C., & Doignon, J.-P. (1988b). A Markovian procedure for assessing the state of a system. *Journal of Mathematical Psychology, 32*, 232–258.

Falmagne, J.-C., Koppen, M., Villano, M., Doignon, J.-P., & Johannesen, L. (1990). Introduction to knowledge spaces: How to build, test, and search them. *Psychological Review, 97*, 201–224.

Falstein, L. D. (1986). *Basic mathematics*. Reading, MA: Addison-Wesley.

Held, T. (1993). *Establishment and empirical validation of problem structures based on domain specific skills and textual properties. A contribution to the "theory of knowledge spaces"*. Unpublished doctoral dissertation, University of Heidelberg, Germany.

Koppen M., & Doignon, J.-P. (1990). How to build a knowledge space by querying an expert. *Journal of Mathematical Psychology, 34*, 311–331.

Korossy, K. (1993). *Modellierung von Wissen als Kompetenz und Performanz. Eine Erweiterung der Wissensstruktur-Theorie von Doignon und Falmagne.* [Modeling knowledge as competence and performance. An extension to Doignon and Falmagne's theory of knowledge spaces.] Unpublished doctoral dissertation, University of Heidelberg, Germany.

Korossy, K. (1996). Kompetenz und Performanz beim Lösen von Geometrie-Aufgaben. [Competence and performance in solving geometry problems.] *Zeitschrift für Experimentelle Psychologie, 43*, 279–318.

Lukas, J., & Albert, D. (1993). Knowledge assessment based on skill assignment and psychological task analysis. In G. Strube & K. F. Wender (Eds.), *The cognitive psychology of knowledge* (pp. 139–159). Amsterdam: Elsevier.

Meserve, B. E., & Sobel, M. A. (1989). *Introduction to mathematics*. Englewood Cliffs, NJ: Prentice-Hall.

Ministerium für Kultus und Sport, Baden-Württemberg (1994). *Bildungsplan für das Gymnasium.* Kultus und Unterricht, Lehrplanheft 4/1994.

Schmid, A. (Ed.) (1994). *Lambacher Schweizer LS 6. Mathematisches Unterrichtswerk für das Gymnasium.* Stuttgart: Ernst Klett.

Schrepp, M. (1993). *Über die Beziehung zwischen kognitiven Prozessen und Wissensräumen beim Problemlösen.* [On the relation between cognitive processes and knowledge spaces in problem solving.] Unpublished doctoral dissertation, University of Heidelberg, Germany.

Setek, W. M. (1992). *Fundamentals of mathematics.* New York: Macmillan.

6

An Empirical Test of a Process Model for Letter Series Completion Problems

Martin Schrepp
Ruprecht-Karls-Universität Heidelberg

The connection between the theory of knowledge spaces and formal models of cognitive problem solving processes is examined. This connection is described for a model of the cognitive processes of subjects in solving letter series completion problems, which is based on ideas from Simon and Kotovsky (1963). The model is formulated as an algorithm, depending on its ability to solve letter series completion problems on two parameters. Different abilities of subjects in solving such problems can be represented within the model by different choices for these parameters. It is shown that the model determines surmise relations on sets of letter series completion problems. These surmise relations, respectively the corresponding quasi–ordinal knowledge spaces, are used in two experimental investigations to test the underlying process model empirically. The results of these experiments show that the surmise relations derived from the process model are able to predict the difficulty of letter series completion problems in a satisfactory manner.

INTRODUCTION

The knowledge of a subject about a special knowledge domain is identified in the theory of knowledge spaces with his or her ability to solve problems from this domain. Such a formalization of human knowledge has the advantage to allow a precise mathematical treatment of this concept, but totally neglects cognitive processes or abilities on which the solution behavior of a subject is based. Therefore, the question arises if it is possible to extend the theory of knowledge spaces developed by Doignon and Falmagne (1985, 1998) in a way that allows the description of such underlying cognitive solution processes.

In this chapter, we examine the connection between the theory of knowledge spaces and process models of the cognitive problem solving process of subjects. We deal especially with the question of how knowledge structures, i.e. sets of knowledge states, can be derived from process models.

A process model describes in detail the cognitive processes subjects use to solve problems. So, it seems possible to derive the set of all solution patterns or knowledge states that are consistent with the assumptions about solution processes contained in the process model. The set of knowledge states consistent with a process model forms the knowledge structure corresponding to that model.

The investigation of the connection between knowledge structure and process model has two applications. First, formal models of the cognitive problem solving process can be used to establish knowledge structures, which can serve as a basis for an efficient assessment of knowledge (see for example Falmagne & Doignon, 1988). Second, process models can be tested empirically through a comparison between the knowledge structures corresponding to them and observed data patterns. We concentrate in this chapter mainly on the second application.

We examine the connection between knowledge structure and process model for a formal model of the cognitive problem solving processes of subjects in the field of letter series completion problems (short lsc-problems). This model, which is especially able to describe interindividual differences in the ability of subjects to solve such lsc-problems, is based on ideas from Simon and Kotovsky (1963).

A general approach to describe the connection between the theory of knowledge spaces and process models can be found in Schrepp (1993).

LETTER SERIES COMPLETION PROBLEMS

In lsc-problems the subjects task is to identify a regularity or rule in a presented sequence of letters of the alphabet and to use this rule to extend the sequence. Such lsc-problems are often used in intelligence tests (e.g., in Thurston's primary mental abilities test) to measure the ability of inductive reasoning.

As an example for a lsc-problem, look at the sequence *axbxcxdx*. What is the rule used to generate this sequence and how can we find a continuation?

First, we see that every letter at an even position in the sequence is an *x*. Second, the letter at the odd position j is the immediate successor in the alphabet of the letter at position $j - 2$. The continuation in accordance to this rule is *exfx*, and so on.

From a strong logical standpoint there exists no unique solution for such lsc-problems. It is, for example, also possible to continue the letter sequence shown above by *axbxcxdx* and so on. This continuation is based on the rule "Repetition of the letter block *axbxcxdx*" and therefore also consistent with the given sequence. But such a continuation is less obvious than the one described above. As Simon and Kotovsky (1963) pointed out, it is easy to find consensus about the "correct" continuation in the lsc-problems commonly used. This correct continuation is in some sense simpler or more natural than other continuations.

To analyse the cognitive processes subjects use to detect regularities in sequences of letters a formal representation of such regularities must be formulated. Such a formal system to represent lsc-problems was introduced by Klahr and Wallace (1970).

The letter sequences of the commonly used lsc-problems are a mixture of simple sequences. In this simple sequences, every letter has a specific relation to his predecessor in the sequence. As an example, the letter sequence *axbxcxdx* consists of the simple sequences *abcd* and *xxxx*. Such a simple sequence can be characterized by the relation that holds between two letters and the first letter in the sequence. Therefore *abcd* can be described by the relation "is next letter in the alphabet" and the first letter *a* and *xxxx* can be described by the relation "is same letter as" and the first letter *x*. This description is, in the following, written by a symbol for the relation followed by the first letter in brackets. So we write $N(a)$ (N for next) to describe the sequence *abcd* and $S(x)$ (S for same) to describe the sequence *xxxx*.

To represent a letter sequence it is sufficient to describe the simple sequences it consists of and the order in which letters from these simple sequences are chosen. This can be done by writing down the relation symbols of the simple sequences in the order of their occurrence in the letter sequence followed by the first letters of the simple sequences in brackets in the same order. We can describe our example *axbxcxdx* by $NS(ax)$. We call such a representation of a letter sequence the *pattern description* of the sequence. This pattern description is, like the continuation of the sequence, not unique. We call the number of relational symbols in the pattern description of a letter sequence the *period* of the sequence.

In this chapter we use also the three relations "is predecessor in the alphabet" (symboliced by P), "is double–next letter in the alphabet" (symboliced by D), and "is triple–next letter in the alphabet" (symboliced by T) to describe letter sequences. As an additional example $NPDT(ekad)$ describes the sequence *ekadfjcggiej*... of letters.

Notice that these relations are defined cyclic concerning the alphabet. We have, for example, $z\,Na$, $a\,Pz$, $y\,Da$, and $x\,Ta$. We call a relational dependency between two letters *cyclic* if these letters lie on different ends of the alphabet. We mark a relation in the pattern description of a lsc-problem by a $*$ if a cyclic dependency must be recognized to continue the sequence correct and call such a relation in the following a *cyclic relation* in this pattern description. For example, the letter sequence *fiehdgcfbead* is characterized by the pattern description P^*P(fi), because for the continuation of this sequence, the cyclic dependency aPz must be recognized.

In the next section we describe the cognitive processes subjects use to fix a pattern description[1] of a lsc-problem.

A MODEL FOR THE SOLUTION PROCESS

In this section, we describe a model of the cognitive processes on which the solution behavior of subjects solving lsc-problems is based. To construct this model we use ideas from the work of Simon and Kotovsky (1963). These authors assume that subjects continue letter sequences by fixing a pattern description of such sequences. This pattern description is then used to determine the correct continuation. The central process in solving lsc-problems is, therefore, the construction of a pattern description for a given letter sequence. Such a pattern description is, accordingly to Simon and Kotovsky (1963), constructed by recognizing relations between specific letters in a letter sequence.

We start with a given letter sequence $x_1 \ldots x_n$. The process of constructing a pattern description for this sequence consists of two subprocesses.

The first subprocess tries to find out the period p of the letter sequence $x_1 \ldots x_n$. For this purpose, it searches for a simple subsequence of $x_1 \ldots x_n$, that means, for a sequence $x_j\ x_{j+p}\ x_{j+2p} \ldots$, in which every letter has a specific relation R to his predecessor in this subsequence. To detect such a subsequence the process uses a number of comparisons between single letters of $x_1 \ldots x_n$. The second subprocess uses the information about the period to detect all the simple subsequences a given letter sequence consists of, i.e. the pattern description of that letter sequence.

The first subprocess starts with comparing for $j = 1, 2, 3, \ldots, k$ the letter x_1 with x_{j+1}. The constant k denotes the largest natural number less than $n/2$. If $x_1 R x_{j+1}$ the process checks if also the dependencies $x_{j+1} R x_{2j+1}$, $x_{2j+1} R x_{3j+1}, \ldots$ holds. If this is the case, then $x_1, x_{j+1}, x_{2j+1}, \ldots$ is a simple subsequence and j is the expected period of the letter sequence. If the first subprocess has up to this point not found such a simple subsequence, it compares

[1] The idea to fix a special representation of regularities underlying letter sequences and to use this representation to describe the cognitive processes of subjects solving lsc-problems is due to Simon and Kotovsky (1963). The representation they used is a little different from the one in Klahr and Wallace (1970), which we described above.

for $j = 2, 3, \ldots, k$ the letter x_2 with x_{j+2}. If $x_2 R x_{j+2}$ holds for a relation R, then the first subprocess checks if $x_{j+2} R x_{2j+2}$, $x_{2j+2} R x_{3j+2}, \ldots$ also holds. If this is the case, then $x_2, x_{j+2}, x_{2j+2}, \ldots$ is a simple subsequence and j the expected period of $x_1, \ldots, x_n$. If the process has up to this point not found a simple subsequence it starts the same comparison process for x_3, and so on.

The second subprocess starts if the first one had successfully detected a simple subsequence and period j. This second subprocess tries to find the additional simple subsequences which are contained in the given letter sequence. This means it searches for relations $R_1, \ldots, R_j$ with:

$$\begin{array}{cccc} x_1 R_1 x_{j+1} \;, & x_{j+1} R_1 x_{2j+1} \;, & x_{2j+1} R_1 x_{3j+1} \;, & \ldots \\ \vdots & \vdots & \vdots & \vdots \\ x_j R_j x_{2j} \;, & x_{2j} R_j x_{3j} \;, & x_{3j} R_j x_{4j} \;, & \ldots \;. \end{array}$$

If the second subprocess is able to find such relations, then $R_1 \ldots R_j(x_1 \ldots x_j)$ is a pattern description of the given letter sequence $x_1 \ldots x_n$. If the second subprocess fails in finding such relations, then the process continues with the first subprocess at the point this process is interrupted.

Up to now, we have only described the basic algorithm. This algorithm is able to detect a pattern description for every letter sequence if it is able to use all necessary relations in the first and respectively second subprocess. To model interindividual differences in the ability of subjects to solve lsc-problems, we have to introduce restrictions about the relations the algorithm can use. These restrictions are formulated by parameters λ_1 and λ_2.

In this chapter, we restrict ourselves to lsc-problems that can be described by using only the relations S, N, P, D, T in their pattern description. But a generalization to lsc-problems that must be described by other relations is straightforward.

Let λ_1 be a 6–tuple $\langle f_S, f_N, f_P, f_D, f_T, f_Z \rangle$, where f_S, f_N, f_P, f_D, f_T are elements of $\{0, \ldots, k\}$ and f_Z is an element of $\{0, 1\}$. We interpret f_R for $R \in \{S, N, P, D, T\}$ as the maximal number of letters in a letter sequence that can lie between x_i and x_j, so that $x_i R x_j$ could be detected by the first subprocess. Therefore, a relation $R \in \{S, N, P, D, T\}$ can be used in the first subprocess only for comparisons $x_i R x_j$ between single letters x_i and x_j if $f_R \geq j - i - 1$. The value f_Z determines the ability of the first subprocess to detect a cyclic relational dependency between two letters. $f_Z = 0$ indicates that the first subprocess can detect only noncyclical relational dependencies. $f_Z = 1$ indicates that there is no such restriction.

For the relations used in the second subprocess, we made a similar assumption. Let λ_2 be a 6–tuple $\langle s_S, s_N, s_P, s_D, s_T, s_Z \rangle$, where s_S, s_N, s_P, s_D, s_T are elements of $\{0, \ldots, k\}$ and s_Z is a element of $\{0, 1\}$. We interpret s_R for $R \in \{S, N, P, D, T\}$ as the maximal number of letters in a letter sequence that can lie between x_i and x_j so that $x_i R x_j$ could be detected by the second subprocess. As in the first subprocess, s_Z determines the ability to detect cyclic relational dependencies.

We assume additionally that $f_S \leq s_S, f_N \leq s_N, f_P \leq s_P, f_D \leq s_D, f_T \leq s_T$, and $f_Z \leq s_Z$. This assumption can be interpreted in the sense that the ability to detect relations between letters increases in the second subprocess. This natural assumption reflects the observation that lsc-problems are much easier to solve if the period has already been detected because the knowledge about the period can be used to search for relations that are, without this information, very hard to find.

It is also very natural to assume $f_S \geq f_N \geq f_P \geq f_D \geq f_T$ and $s_S \geq s_N \geq s_P \geq s_D \geq s_T$. This assumption reflects the observation that the relations themselves could be ordered in respect to the difficulty to detect them between single letters. For example, $f_S \geq f_N$ and $s_S \geq s_N$ means that it is easier to detect the relation S than to detect the relation N.

The ability of the described algorithm to construct pattern descriptions of letter sequences depends on the two parameters λ_1 and λ_2. We demonstrate this dependency by an example. Define $\tau_1 := \langle 2, 0, 0, 0, 0, 0\rangle$, $\tau_2 := \langle 2, 2, 0, 0, 0, 0\rangle$ and $\sigma_1, \sigma_2 := \langle 2, 1, 0, 0, 0, 0\rangle$. For the choice $\lambda_1 = \tau_1$ and $\lambda_2 = \tau_2$ the algorithm is able to construct the pattern description $SNS(urt)$ of the letter sequence *urtustutt* and fails in constructing the pattern description $NN(ad)$ of the letter sequence *adbecfdgeh* (because $f_N = 0$ indicates that the first subprocess can not detect the relation N). For the choice $\lambda_1 = \sigma_1$ and $\lambda_2 = \sigma_2$ the converse holds. The algorithm can construct $NN(ad)$ and fails in constructing $SNS(urt)$ (because $s_N = 1$ indicates that the second subprocess is not able to detect the relation $x_i N x_j$ if $j - i \geq 2$).

We interpret the algorithm for each choice of the two parameters λ_1 and λ_2 as a model for the cognitive processes of a specific subject in constructing pattern descriptions of letter sequences. So, a subject is in our model represented by a tuple $\langle \lambda_1, \lambda_2 \rangle$ of parameters. Our last two assumptions concerning the values f_R and s_R for $R \in \{S, N, P, D, T, Z\}$ restrict the possible parameter combinations $\langle \lambda_1, \lambda_2 \rangle$.

Formulations of this model and two similar models as production systems can be found in Schrepp (1993).

A detailed description of the algorithm is given by the flow chart in Fig. 1.

IMPLICATIONS ON THE DIFFICULTY OF LETTER SERIES COMPLETION PROBLEMS

The algorithm introduced in the previous section together with our assumptions concerning the possible parameter combinations can be used to derive a surmise relation $\sqsubseteq$ on a set of lsc-problems Q. We use the surmise relation $\sqsubseteq$ and the corresponding quasi-ordinal knowledge space $\mathcal{W}_{\sqsubseteq}$ in the next section for an em-

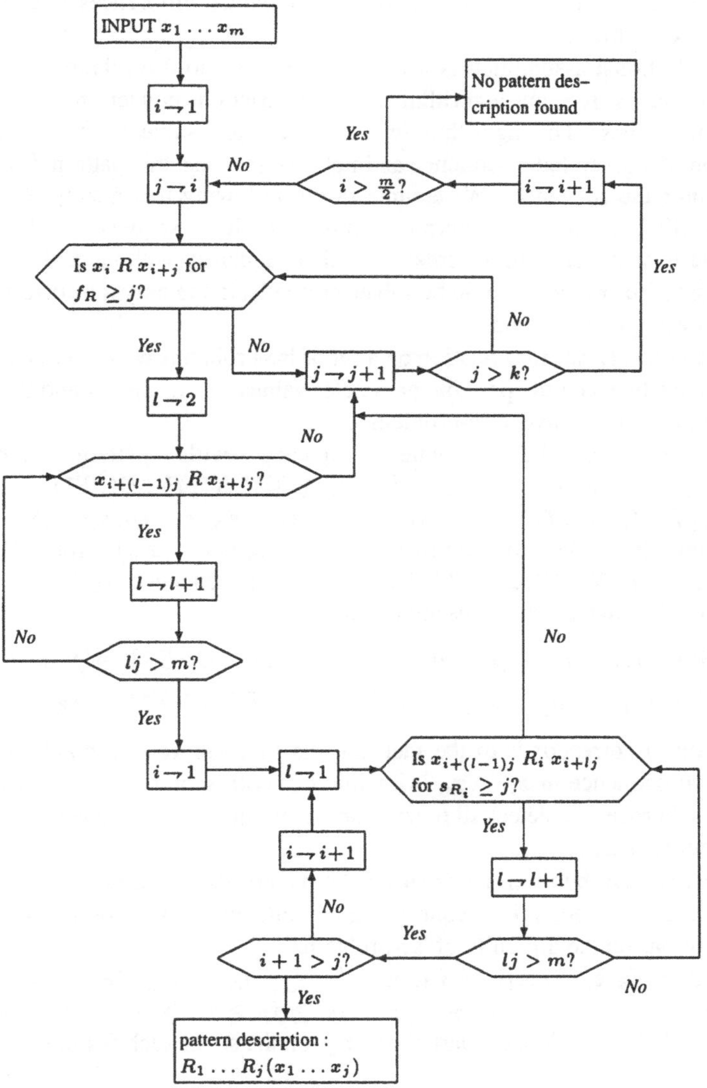

FIG. 1. Flow chart of the algorithm for the construction of a pattern description for a given letter sequence $x_1 \ldots x_n$. The symbol $\rightharpoondown$ should be interpreted as assigning a value to a variable, so $l \rightharpoondown l+1$ means that the new value of the variable l is given by the old value of that variable plus one.

pirical test of our approach to describe interindividual differences of subjects in solving lsc-problems.

An additional assumption is needed to derive a surmise relation on a set Q of lsc-problems from our algorithm and our restrictions concerning the possible parameter values. The algorithm describes only how subjects fix a pattern description of a given letter sequence and not how they use this pattern description to continue the sequence. We assume in the following that a subject who has successfully constructed a pattern description of a letter sequence is also able to continue this sequence in accordance to that regularity without making errors. Therefore a lsc-problem should be solved correctly, if and only if a corresponding pattern description is found.

To construct a surmise relation on a set of lsc-problems we have to clarify the dependency between the possible parameter values for λ_1 and λ_2 and the ability of the algorithm to solve a lsc-problem.

Let $x_1 \ldots x_n$ be a letter sequence with pattern description $R_1 \ldots R_m(x_1 \ldots x_m)$, $R \in \{S, N, P, D, T, N^*, P^*, D^*, T^*\}$, $\lambda_1 = \langle f_S, f_N, f_P, f_D, f_T, f_Z \rangle$ and $\lambda_2 = \langle s_S, s_N, s_P, s_D, s_T, s_Z \rangle$. The algorithm is able to construct $R_1 \ldots R_m(x_1 \ldots x_m)$ for $R_i \in \{S, N, P, D, T, N^*, P^*, D^*, T^*\}$ from the given sequence $x_1 \ldots x_n$, if and only if the following conditions are fulfilled:

a) $\exists i \in \{1, \ldots, m\} \; (f_{R_i} \geq m \wedge (R_i \in \{N^*, P^*, D^*, T^*\} \rightarrow f_Z = 1))$

b) $\forall i \in \{1, \ldots, m\} \; (s_{R_i} \geq m \wedge (R_i \in \{N^*, P^*, D^*, T^*\} \rightarrow s_Z = 1))$.

Condition a) corresponds to the fact that the first subprocess is able to find a simple subsequence in $x_1 \ldots x_n$. Condition b) corresponds to the ability of the second subprocess to detect all other simple subsequences contained in the given letter sequence $x_1 \ldots x_n$.

Now, we can build up an order on the pattern descriptions of lsc-problems that reflects our assumptions concerning the influence of the possible parameter combinations on the difficulty of lsc-problems.

Let $\sigma = \langle i_S, i_N, i_P, i_D, i_T, i_Z \rangle$ and $\sigma' = \langle j_S, j_N, j_P, j_D, j_T, j_Z \rangle$ be 6–tuples with $i_S \geq i_N \geq i_P \geq i_D \geq i_T$, $j_S \geq j_N \geq j_P \geq j_D \geq j_T \in \{0, \ldots, k\}$, and $i_Z, j_Z \in \{0, 1\}$. We define a partial order $\preceq$ on the set of such 6–tuples by :

$$\sigma \preceq \sigma' \Leftrightarrow i_S \leq j_S \wedge i_N \leq j_N \wedge i_P \leq j_P \wedge i_D \leq j_D \wedge i_T \leq j_T \wedge i_Z \leq j_Z.$$

Assume a letter sequence $x_1 \ldots x_n$. Let $\sigma_1(x_1 \ldots x_n)$ be the smallest choice for the parameter λ_1 with respect to $\preceq$, so that the first subprocess of the algorithm can find a simple subsequence in $x_1 \ldots x_n$. Let $\sigma_2(x_1 \ldots x_n)$ the smallest choice for the parameter λ_2 with respect to $\preceq$, so that the second subprocess of the algorithm can find all other simple subsequences contained in $x_1 \ldots x_n$. It is clear that such smallest choices $\sigma_1(x_1, \ldots, x_n)$ and $\sigma_2(x_1 \ldots x_n)$ concerning $\preceq$ for these parameters exist in every case. Note that $\langle \sigma_1(x_1 \ldots x_n), \sigma_2(x_1 \ldots x_n) \rangle$ is a possible parameter combination accordingly to our assumptions concerning the values f_R and s_R for $R \in \{S, N, P, D, T\}$. As an example for the letter sequence x_1 :=*axbxcxdx* we have $\sigma_1(x_1) = \langle 1,0,0,0,0,0\rangle$ and $\sigma_2(x_1) = \langle 1,1,0,0,0,0\rangle$ and for the letter sequence x_2 :=*aobncmdl* we have $\sigma_1(x_2) = \langle 1,1,0,0,0,0\rangle$ and $\sigma_2(x_2) = \langle 1,1,1,0,0,0\rangle$.

Now we are able to order lsc-problems with respect to their difficulty by defining for letter sequences $x := x_1 \ldots x_n$ and $y := y_1 \ldots y_m$ a relation $\sqsubseteq^*$ through the condition :

$$x \sqsubseteq^* y \Leftrightarrow \sigma_1(x) \preceq \sigma_1(y) \wedge \sigma_2(x) \preceq \sigma_2(y).$$

So for the two examples x_1 and x_2 we have $x_1 \sqsubseteq^* x_2$ because $\sigma_1(x_1) = \langle 1,0,0,0,0,0\rangle \preceq \sigma_1(x_2) = \langle 1,1,0,0,0,0\rangle$ and $\sigma_2(x_1) = \langle 1,1,0,0,0,0\rangle \preceq \sigma_2(x_2) = \langle 1,1,1,0,0,0\rangle$.

The relation $\sqsubseteq^*$ is as a conjunction of two quasi–orders [2] also a quasi–order. It reflects our assumptions concerning the dependency between possible parameter combinations and the difficulty of lsc-problems.

Assume $x \sqsubseteq^* y$. The algorithm is for the choices $\lambda_1 = \sigma_1(y)$ and $\lambda_2 = \sigma_2(y)$ able to construct the pattern description of y. But then the algorithm is also able to construct the pattern description of x since $\sigma_1(x) \preceq \sigma_1(y)$ and $\sigma_2(x) \preceq \sigma_2(y)$. Because $\lambda_1 = \sigma_1(y)$ and $\lambda_2 = \sigma_2(y)$ are minimal with respect to $\preceq$ the algorithm is for any choice of the parameters able to solve x if he is able to solve y. Therefore $x \sqsubseteq^* y$ for two lsc-problems x and y can be interpreted as "If the algorithm is for a choice of values for λ_1 and λ_2 able to construct the pattern description of y then he is also able to construct the pattern description of x".

A problem is that according to $\sqsubseteq^*$ only the relations in the pattern description of a letter sequence influences the difficulty to construct a pattern description of this sequence. As an example, look at the letter sequences *aabccedg* and *agbickdm*. Their pattern descriptions are $ND(aa)$ and $ND(ag)$ and therefore with respect to the relations in these pattern descriptions identical. Therefore *aabccedg* $\sqsubseteq^*$ *agbickdm* and *agbickdm* $\sqsubseteq^*$ *aabccedg*. However intuitively it seems much harder to continue the first sequence than the second. The reason is that in the first sequence a lot of relations between letters come in mind that are superfluous for the construction of the pattern description and therefore maybe confusing. For example, in this sequence the first two letters are identical and the third letter is the alphabetical successor of the second letter.

Our algorithm is already able to reflect this intuitively expected difference in difficulty between this examples, since he needs more processing steps to construct a pattern description of *aabccedg* than to construct a pattern description of *agbickdm*. The reason is that the first subprocess detects in solving *aabccedg* the

[2] Note that the generalization of the partial order $\preceq$ from the set of parameters to the set of letter sequences gives only a quasi–order because different letter sequences may correspond to the same values for σ_1 and σ_2.

relation S between the first and the second letter. Therefore, it had to find out that this relation does not hold between the second letter and the third. Only after this two steps can the correct dependency N between the first and third letter be detected. In contrast the algorithm detects no relation between first and second letter of *agbickdm*. The correct relation N between the first and the third letter can, therefore, be detected directly. Thus, constructing the pattern description of *aabccedg* needs more processing steps (look at the flow chart in Fig. 1) than constructing the pattern description of *agbickdm*.

The algorithm is able to construct for any parameter combination either a pattern description for both problems or for none of them, but a different number of processing steps is needed in both cases to do so. If we assume a monotonic dependency between processing steps needed to construct a pattern description and solution time the solution of *aabccedg* should take more time than the solution of *agbickdm*. In situations in which the solution time is restricted (as in the experiments we report in the next section) it is therefore consistent with our model that a solution of *agbickdm* is given while no such solution of *aabccedg* can be found.

So, we have to change $\sqsubseteq^*$ in a way that takes into account the number of steps our algorithm needs for constructing the pattern descriptions of letter sequences. This is relatively simple because such differences in the necessary processing steps can only result from detected relations between letters that are not in accordance with the pattern description. For example, in the sequence *aabccedg* the relation S is detected between the first two letters, but this relation is not in accordance with the pattern description $ND(\mathrm{aa})$ of this sequence.

We define a surmise relation $\sqsubseteq$ by $x \sqsubseteq y$ if and only if $x \sqsubseteq^* y$ and for constructing the pattern description of x at least as much processing steps are necessary than for the construction of the pattern description of y[3]. For example, we have *agbickdm* $\sqsubseteq$ *aabccedg* but *aabccedg* $\not\sqsubseteq$ *agbickdm*. To find out if $x \sqsubseteq y$ holds or not we have for pairs x, y of lsc-problems with $x \sqsubseteq^* y$ only to count how often the algorithm detects a relation that is not in accordance with the finally constructed pattern description[4].

EXPERIMENT I

To test our model empirically, a problem set containing 20 lsc-problems was chosen and the surmise relation $\sqsubseteq_I$ based on the principles described in the last section was constructed. We compare $\sqsubseteq_I$ and the corresponding quasi–ordinal knowledge space $\mathcal{W}_{\sqsubseteq_I}$ with the obtained data patterns of 51 subjects.

[3] Note that this procedure for constructing $\sqsubseteq$ from $\sqsubseteq^*$ can not yield intransitivities, because if we have $x \sqsubseteq^* y \sqsubseteq^* z$ and $x \not\sqsubseteq z$ we have also $x \not\sqsubseteq y$ or $y \not\sqsubseteq z$. $\sqsubseteq$ is therefore transitive.

[4] The length of the period of a pattern sequence influences also the number of processing steps needed by the algorithm to construct a pattern description. However, because $x \sqsubseteq^* y$ implies that the period of y is greater or equal to the period of x, we can neglect this influence.

TABLE 1
Lsc–Problems Used in Experiment I.

problem number	*letter sequence*	*pattern description*	*correct continuation*
1	*cdcdcdcd*	$SS(cd)$	*cdc*
2	*gohpiqjrks*	$NN(go)$	*ltm*
3	*tbaxtbaxtb*	$SSSS(tbax)$	*axt*
4	*fiehdgcfbe*	$P^{*}P(fi)$	*adz*
5	*abyabxabwab*	$SSP(aby)$	*vab*
6	*adbecfdgeh*	$NN(ad)$	*fig*
7	*jkorklpslmqt*	$NNNN(jkor)$	*mnr*
8	*ehgjilknmp*	$DD(eh)$	*orq*
9	*urtustuttu*	$SNS(urt)$	*utu*
10	*pvountmslr*	$PP(pv)$	*kqj*
11	*eafgbhicjkdl*	$DND(eaf)$	*men*
12	*dhefcgdebfcd*	$PPPP(dhef)$	*aeb*
13	*npaoqapraqsa*	$NNS(npa)$	*rta*
14	*axdcxfexhgxj*	$DSD(axd)$	*ixl*
15	*ekofkngkmhk*	$NSP(eko)$	*lik*
16	*jkillminnoip*	$DDSD(jkil)$	*pqi*
17	*fgjggkhglig*	$NSN(fgj)$	*mjg*
18	*wrnfvqmeupld*	$PPPP(wrnf)$	*tok*
19	*dhcbfjedhlgf*	$DDDD(dhcb)$	*jni*
20	*mrnorpqrrsrt*	$DSD(mrn)$	*urv*

Method

Subjects

The experiment was conducted with 25 female and 26 male subjects. Their ages ranged from 12 to 53 years. The average age was 26 years. For their participation, the subjects were paid with 12,– DM. The subjects were recruited by an announcement in the local newspaper and the institutes of the university.

Problems

The problem set consists of 20 lsc-problems that are shown in Table 1. We refer to this problem set in the following as Q_I. Some of these problems were constructed especially for the investigation; others were taken from the literature concerning lsc-problems. In each problem, three letters has to be given as a continuation of the sequence by the subjects. Note that in Problem 4 a cyclic relational dependency has to be detected to continue the sequence.

Derived surmise relation

In Fig. 2 the surmise relation $\sqsubseteq_I$ on the problem set Q_I is shown as a Hasse–Diagram. The corresponding quasi-ordinal knowledge space $\mathcal{W}_{\sqsubseteq_I}$ consists of 276 knowledge states.

Procedure

The investigation was conducted as a group experiment. The size of a group varied between 3 and 11 subjects. First, a general instruction was handed out that explained the problem type "lsc-problems". Then the subjects had to solve 4 lsc-problems as exercises. After all subjects of a group had finished these exercises, they were informed about the correct solutions.

Then, a second instruction was handed out that informed the subjects about the general experimental conditions. The subjects were instructed not to spend more than 2 minutes on each problem. They were able to control this time themselves on a watch in the room.

After all subjects had finished reading this second instruction the 20 lsc-problems from Table 1 were handed out. Each problem was printed on a separate sheet. The problems were ordered in increasing sequence of their numbers in Table 1. The subjects were instructed to work on the problems in exactly this order and not to go back to problems already finished. As solution they had to write

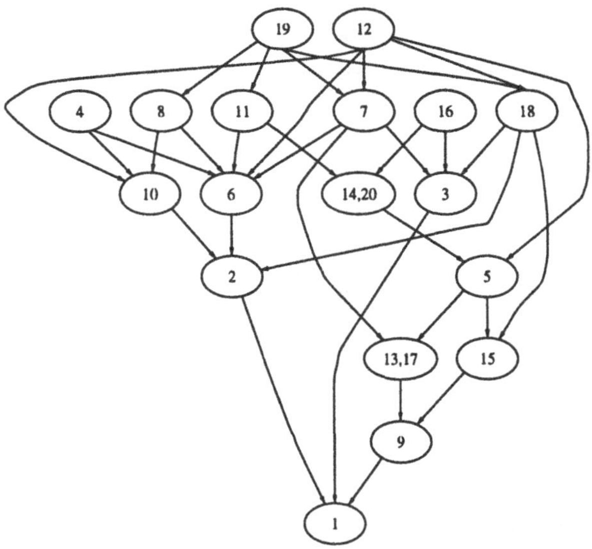

FIG. 2. Hasse–Diagram of the derived surmise relation $\sqsubseteq_I$ on the problem set Q_I.

TABLE 2
Frequency of Symmetric Distances Between Response Patterns
and Closest States of the Quasi-Ordinal Knowledge Space $\mathcal{W}_{\sqsubseteq_I}$ Corresponding to $\sqsubseteq_I$.

	Distances						*Average*
	0	*1*	*2*	*3*	*4*	*5*	*Distance*
Frequencies	23	14	9	5	0	0	0.92

down three letters as a continuation behind the presented sequence. Detailed informations concerning the experimental procedure of Experiment I can be found in Schrepp (1993).

Results

The average solution frequency was 81 %. This high average solution frequency results mainly from the fact that some problems (1, 2, 3, 5, 9, 10, 11, 13, 15) were solved by a large majority (more than 46) of subjects.

First we compare the observed response patterns with the knowledge space $\mathcal{W}_{\sqsubseteq_I}$. Therefore we look at the symmetric distances [5] between response patterns and the closest states[6] of $\mathcal{W}_{\sqsubseteq_I}$ shown in Table 2.

Notice that because $\mathcal{W}_{\sqsubseteq_I}$ includes the whole problem set Q_I and $\emptyset$ the highest theoretically possible symmetric distance between a response pattern and a corresponding closest state in $\mathcal{W}_I$ is 10. Because of the high average solution frequency, the highest possible average symmetric distance of an arbitrary knowledge structure containing Q_I to our data set is 3.8.

To get more information, about the deviations between our data and the theoretically derived surmise relation $\sqsubseteq_I$, we counted for each pair a, b of problems from Q_I with $a \sqsubseteq_I b$ how often this assertion is violated empirically. A violation of $a \sqsubseteq_I b$ means that b is solved by a subject that fails in solving a. Table 3 shows for each pair a, b of lsc-problems with $a \sqsubseteq_I b$ the number of such violations.

Discussion

Table 2 shows that the symmetric distances between observed response patterns of subjects and the closest states of $\mathcal{W}_{\sqsubseteq_I}$ are rather small. Note that 23 response

[5] The symmetric distance d between two subsets A and B of Q is defined by $d(A, B) := | A \triangle B | = | (A \setminus B) \cup (B \setminus A) |$.

[6] A closest state in a knowledge structure $\mathcal{W}$ for an observed response pattern R is a knowledge state W of $\mathcal{W}$ with $d(W, R) = \min\{(d(W', R) \mid W' \in \mathcal{W}\}$. Notice that for an observed response pattern there are in general many closest states in a knowledge structure.

TABLE 3
Violations of Relational Dependencies.
The i–th Row and the j–th Column of the Table Contains an – if $i \not\sqsubseteq_I j$.
If $i \sqsubseteq_I j$ the Number of Observed Violations of this Dependency is Shown.

	1	2	3	4	5	6	7	8	9	10	11	12	13	14	15	16	17	18	19	20
1	0	0	0	0	0	0	0	0	0	0	0	0	0	0	0	0	0	0	0	0
2	-	0	-	2	-	3	0	1	-	4	3	1	-	-	-	-	-	4	0	-
3	-	-	0	-	-	-	0	-	-	-	-	0	-	-	-	0	-	0	0	-
4	-	-	-	0	-	-	-	-	-	-	-	-	-	-	-	-	-	-	-	-
5	-	-	-	-	0	-	-	-	-	-	0	0	-	0	-	0	-	-	0	0
6	-	-	-	4	-	0	2	3	-	-	4	0	-	-	-	-	-	-	0	-
7	-	-	-	-	-	-	0	-	-	-	-	3	-	-	-	-	-	-	3	-
8	-	-	-	-	-	-	-	0	-	-	-	-	-	-	-	-	-	-	0	-
9	-	-	-	-	2	-	2	-	0	-	2	1	2	2	2	2	2	2	1	2
10	-	-	-	1	-	-	-	1	-	0	-	0	-	-	-	-	-	-	0	-
11	-	-	-	-	-	-	-	-	-	-	0	-	-	-	-	-	-	-	0	-
12	-	-	-	-	-	-	-	-	-	-	-	0	-	-	-	-	-	-	-	-
13	-	-	-	-	4	-	1	-	-	-	3	0	0	3	-	2	2	-	0	3
14	-	-	-	-	-	-	-	-	-	-	6	-	-	0	-	6	-	-	0	5
15	-	-	-	-	3	-	-	-	-	-	2	0	-	1	0	2	-	1	1	2
16	-	-	-	-	-	-	-	-	-	-	-	-	-	-	-	0	-	-	-	-
17	-	-	-	-	6	-	4	-	-	-	6	1	5	5	-	6	0	-	1	5
18	-	-	-	-	-	-	-	-	-	-	-	1	-	-	-	-	-	0	1	-
19	-	-	-	-	-	-	-	-	-	-	-	-	-	-	-	-	-	-	0	-
20	-	-	-	-	-	-	-	-	-	-	6	-	-	6	-	6	-	-	2	0

patterns (45 %) are states of $\mathcal{W}_{\sqsubseteq_I}$ whereas $\mathcal{W}_{\sqsubseteq_I}$ contains only 276 (0.00026 %) of the 2^{20} possible response patterns.

This is a relatively satisfactory result because we have to consider influences like, for example, lucky guesses, careless errors, fatigue or motivational decrease[7] during the experimental procedure.

The number of violations of $i \sqsubseteq_I j$ in Table 3 is for all such pairs of problems less or equal to 6 (of 51 subjects). In conclusion, the assumptions of $\sqsubseteq_I$ concerning the difficulty of lsc-problems seems to be in accordance with the data.

[7] If we assume for example that the probability of both careless errors and lucky guesses is 0.05, then the probability that the response pattern of a subject is equal to his knowledge state is $0.95^{20} \approx 0.35$.

EXPERIMENT II

In our second experiment, a set Q_{II} containing 24 lsc-problems was chosen and the surmise relation $\sqsubseteq_{II}$ on Q_{II} based on the principles derivated from the process model was derived. We compare $\sqsubseteq_{II}$ and the corresponding quasi–ordinal knowledge space $\mathcal{W}_{\sqsubseteq_{II}}$ with the observed data patterns of 53 subjects.

Method

Subjects

The experiment was conducted with 26 female and 27 male subjects. Their ages ranged from 15 to 64 years. The average age was 26 years. For their participation, the subjects were paid with 12,– DM. The subjects were recruited by an announcement in the local newspaper and the institutes of the University of Heidelberg.

Problems

The problem set Q_{II} consists of 24 lsc-problems which are shown in Table 4. In each problem, the next four letters has to be given as a continuation of the sequence by the subjects. In the problems 5, 20, 22, a cyclic relational dependency has to be detected to continue the sequence.

Derived Surmise Relation

In Fig. 3, the surmise relation $\sqsubseteq_{II}$ is depicted as a Hasse–Diagram. The corresponding quasi–ordinal knowledge space $\mathcal{W}_{\sqsubseteq_{II}}$ consists of 2942 knowledge states.

Procedure

The investigation was conducted as a single case experiment that was completely controlled by a computer. The experimental program was developed using MEL (see Schneider, 1990).

First, the subjects received some general informations about the problem-type "lsc-problems". This information was followed by instructions about the way in which the problems would be presented on the screen and the usage of the keyboard to type in the solutions. Then, five lsc-problems are presented as an exercise. This exercise should enable the subject to become familiar with lsc-problems and to train the experimental procedure. Feedback about the correct solution was given after each of the five exercise problems.

After a subject had finished the exercises, a second instruction was presented on the screen. The subject was informed in this instruction that he or she had 3 minutes time to solve each of the following 24 problems. The remaining time for a problem was shown in seconds on the screen counting from 180 down to 0. If a subject exceeded the maximal time, the problem presentation was terminated automatically and the problem was counted as unsolved.

Then, the 24 problems from Table 4 were presented. The presentation of a problem was terminated as soon as the subject typed in a continnuation (or exceeded the time limit). The subject was able to start the presentation of the next problem by himself through pressing one of the function keys. This enabled the subject to take small breaks between two problems. The order of problem presentation was randomized for each subject. During this experimental phase, no feedback about the correctness of the solutions was presented. For detailed informations concerning this experiment see Schrepp (1993).

TABLE 4
Lsc–Problems Used in Experiment II.

problem number	*letter sequence*	*pattern description*	*correct continuation*
1	*bhibhibhibhi*	SSS(*bhi*)	*bhib*
2	*blqcmrdnseot*	NNN(*blq*)	*fpug*
3	*lokojoiohogo*	PS(*lo*)	*foeo*
4	*cmjeolgqnisp*	DDD(*cmj*)	*kurm*
5	*apzpypxpwpvp*	P^*S(*ap*)	*uptp*
6	*wrnvqmupltok*	PPP(*wrn*)	*snjr*
7	*jkjkjilihmhg*	NPP(*jkj*)	*ngfo*
8	*dlngoqjrtmuw*	TTT(*dln*)	*pxzs*
9	*ckofdlpgemqh*	$NNNN$(*ckof*)	*fnri*
10	*difkhmjolqns*	DD(*di*)	*purw*
11	*jknllmnnnonp*	$DDSD$(*jknl*)	*pqnr*
12	*blcmdneofpgq*	NN(*bl*)	*hris*
13	*lpdjnrflpthn*	$DDDD$(*lpdj*)	*rvjp*
14	*eofngmhlikjj*	NP(*eo*)	*kilh*
15	*hhigjfkeldmc*	NP(*hh*)	*nboa*
16	*pvountmslrkq*	PP(*pv*)	*jpio*
17	*jcxglexingxk*	$DDSD$(*jcxg*)	*pixm*
18	*falxiboxlcrx*	$TNTS$(*falx*)	*odux*
19	*wrnfvqmeupld*	$PPPP$(*wrnf*)	*tokc*
20	*ludknwdmpydo*	DD^*SD(*ludk*)	*radq*
21	*emrflqgkphjo*	NPP(*emr*)	*iinj*
22	*jfkeldmcnboa*	NP^*(*jf*)	*pzqy*
23	*lagcmdfengeg*	$NTPD$(*lagc*)	*ojdi*
24	*ekofkngkmhkl*	NSP(*eko*)	*ikkj*

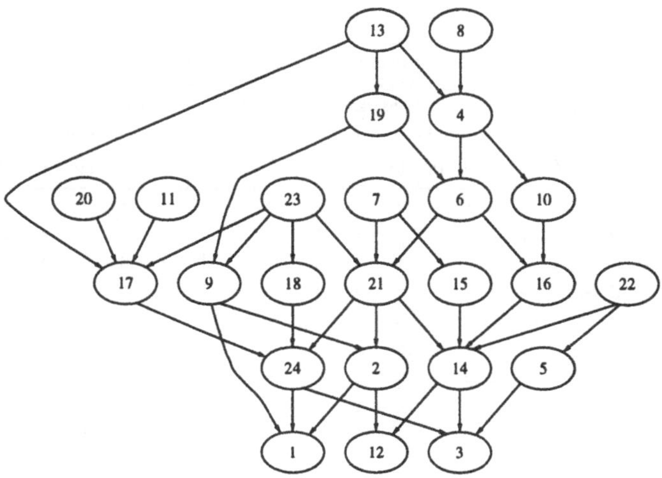

FIG. 3. Hasse–Diagram of the derived surmise relation $\sqsubseteq_{II}$ on the problem set Q_{II}.

TABLE 5
Frequency of Symmetric Distances Between Response Patterns and Closest States of the Quasi-Ordinal Knowledge Space $\mathcal{W}_{\sqsubseteq_{II}}$ Corresponding to $\sqsubseteq_{II}$.

Distances							*Average*
0	*1*	*2*	*3*	*4*	*5*	*6*	*Distance*
14	18	12	4	3	2	0	1.43

Results

The average solution frequency over all subjects and all problems was 79 %. As in Experiment I this high average solution frequency results mainly from the fact that some of the problems were to easy for a large majority of subjects participating in the experimental investigation. The problems 1, 2, 3, 5, 9, 12, 14, 15 were solved by more than 46 subjects.

Table 5 shows the frequency of symmetric distances between the observed response patterns and the closest states of the quasi–ordinal knowledge space $\mathcal{W}_{\sqsubseteq_{II}}$. Because $\mathcal{W}_{\sqsubseteq_{II}}$ includes the whole problem set Q_{II} and $\emptyset$, the highest theoretically possible symmetric distance between an observed response pattern and a corresponding closest state in $\mathcal{W}_{II}$ is 12. Because of the high average solution frequency, the highest possible average symmetric distance of an arbitrary knowledge structure including Q_{II} to our data set is 5.04.

The number of violations of assertions of the form $a \sqsubseteq_{II} b$ is for each such pair a, b of problems from Table 4 listed in Table 6.

TABLE 6
Violations of Relational Dependencies.
The i–th Row and the j–th Column of The Table Contains an - if $i \not\sqsubseteq_{II} j$.
If $i \sqsubseteq_{II} j$ the Number of Observed Violations of This Dependency is Shown.

	1	2	3	4	5	6	7	8	9	10	11	12	13	14	15	16	17	18	19	20	21	22	23	24
1	0	1	-	1	-	1	1	1	1	-	-	-	1	-	-	-	-	-	1	-	1	-	1	1
2	-	0	-	0	-	1	0	1	2	-	-	-	1	-	-	-	-	-	2	-	2	-	2	-
3	-	-	0	0	2	1	0	0	-	0	0	-	0	0	2	1	1	0	1	0	0	1	0	1
4	-	-	-	0	-	-	-	5	-	-	-	-	4	-	-	-	-	-	-	-	-	-	-	-
5	-	-	-	-	0	-	-	-	-	-	-	-	-	-	-	-	-	-	-	-	-	3	-	-
6	-	-	-	3	-	0	-	2	-	-	-	-	3	-	-	-	-	-	3	-	-	-	-	-
7	-	-	-	-	-	-	0	-	-	-	-	-	-	-	-	-	-	-	-	-	-	-	-	-
8	-	-	-	-	-	-	-	0	-	-	-	-	-	-	-	-	-	-	-	-	-	-	-	-
9	-	-	-	-	-	-	-	-	0	-	-	-	0	-	-	-	-	-	1	-	-	-	0	-
10	-	-	-	7	-	-	-	6	-	0	-	-	6	-	-	-	-	-	-	-	-	-	-	-
11	-	-	-	-	-	-	-	-	-	-	0	-	-	-	-	-	-	-	-	-	-	-	-	-
12	-	1	-	0	-	1	0	0	1	0	-	0	0	0	0	0	-	-	0	-	1	0	0	-
13	-	-	-	-	-	-	-	-	-	-	-	-	0	-	-	-	-	-	-	-	-	-	-	-
14	-	-	-	1	-	4	1	1	-	0	-	-	1	0	5	2	-	-	2	-	1	5	0	-
15	-	-	-	-	-	-	1	-	-	-	-	-	-	-	0	-	-	-	-	-	-	-	-	-
16	-	-	-	5	-	9	-	4	-	5	-	-	3	-	-	0	-	-	4	-	-	-	-	-
17	-	-	-	-	-	-	-	-	-	-	6	-	6	-	-	-	0	-	-	-	-	-	8	-
18	-	-	-	-	-	-	-	-	-	-	-	-	-	-	-	-	-	0	-	-	-	-	1	-
19	-	-	-	-	-	-	-	-	-	-	-	-	2	-	-	-	-	-	0	-	-	-	-	-
20	-	-	-	-	-	-	-	-	-	-	-	-	-	-	-	-	-	-	-	0	-	-	-	-
21	-	-	-	6	-	10	3	6	-	-	-	-	4	-	-	-	-	-	9	-	0	-	4	-
22	-	-	-	-	-	-	-	-	-	-	-	-	-	-	-	-	-	-	-	-	-	0	-	-
23	-	-	-	-	-	-	-	-	-	-	-	-	-	-	-	-	-	-	-	-	-	-	0	-
24	-	-	-	3	-	5	2	2	-	-	0	-	2	-	-	-	3	5	4	3	1	-	1	0

Discussion

As Table 5 shows, the symmetric distances between observed response patterns and the closest states of $\mathcal{W}_{\sqsubseteq_{II}}$ are, in almost all cases, rather small. 14 of the observed response patterns (26 %) are states of $\mathcal{W}_{\sqsubseteq_{II}}$ while $\mathcal{W}_{\sqsubseteq_{II}}$ contains only 2942 (0.00017 %) of the 2^{24} possible response patterns.

The observed average distance between response patterns and closest states in Experiment II is higher than in Experiment I. This is not a surprise for two reasons. First, $\mathcal{W}_{\sqsubseteq_{II}}$ is smaller than $\mathcal{W}_{\sqsubseteq_{I}}$ in the sense that it contains a smaller percentage of the possible response patterns (0.00017 % compared to 0.00026 %) and therefore we expect a higher average distance. Second, the influence of lucky guesses, careless errors, fatigue, or motivational decrease on the resulting distances will clearly increase with the number of problems presented. Because we have presented more problems in Experiment II than in Experiment I (24 compared to 20), a higher average distance should be observed in Experiment II.

The observed number of violations of relational dependencies is, as Table 6 shows, with a few exceptions ($10 \sqsubseteq 4, 16 \sqsubseteq 6, 21 \sqsubseteq 6, 21 \sqsubseteq 19, 17 \sqsubseteq 23$) rather small (less than 7 of 53 subjects). The assumptions about the difficulty of lsc-problems contained in $\sqsubseteq_{II}$ seems to be in accordance with the data.

GENERAL DISCUSSION

We examined the connection between knowledge structures and models of the cognitive problem solving process for the example of a model for cognitive processes in letter series completion. We demonstrated how knowledge structures, in our special case quasi-ordinal knowledge spaces corresponding to surmise relations, can be derived from this process model and how these derived knowledge structures can be used for an empirical test of the model.

In the derivation of the surmise relation $\sqsubseteq$ from our process model we can distinguish two steps. First, we derived a quasi–order $\sqsubseteq^*$ from the model using only the information concerning the relations in the pattern descriptions of lsc-problems. This first step can be completely characterized as a special skill assignment (see Doignon, 1994). For the construction of a knowledge structure by a skill assignment a set of cognitive skills that are important in the underlying knowledge domain is fixed. The skills are then used to derive the set of all possible knowledge states of subjects by postulating a connection between the ability to solve problems from the knowledge domain and the mastery of these skills. In our special case, the set of skills is given by the set of all possible parameter tuples $\langle \lambda_1, \lambda_2 \rangle$ and the connection between skills and the ability to solve lsc-problems is given by the conditions a) and b). In a second step, we used detailed information about the processing steps of the algorithm to construct the final surmise relation $\sqsubseteq$ from $\sqsubseteq^*$. The detailed description of the solution process contained in the process model is not used until this second step.

In the following, we discuss if the reported experimental investigations support the described model for cognitive processes in letter series completion. In both experiments, the symmetric distances between the observed response patterns and the corresponding closest states in the quasi–ordinal knowledge spaces derived from the process model were rather small. These observed symmetric distances can be explained by influences like lucky guesses, careless errors or motivational decrease during the experimental procedure.

An additional analysis of the predicted relational dependencies between lsc-problems shows that they are violated empirically only in a small number of cases, with a few exceptions in Experiment II. The assumptions about the difficulty of lsc-problems contained in the surmise relations derived from the process model seems to be in good accordance with the data observed in the two experiments, even if we take into account the high average solution frequency in these experiments.

But, we have to mention also that the surmise relations $\sqsubseteq_I$ and $\sqsubseteq_{II}$ seems to be not complete, that means there seems to be more relational dependencies between lsc-problems than predicted by these surmise relations. The surmise relations $\sqsubseteq$ on sets of lsc-problems derived from our process model depends strongly on the assumptions about the possible parameter tuples described previously. We have formulated in this section only those restrictions that seem to be very natural in respect to our interpretation of these parameters. Additional restrictions concerning the possible parameter combinations will lead to stricter surmise relation $\sqsubseteq'$. Such a surmise relation $\sqsubseteq'$ will contain $\sqsubseteq$ but will in general predict additional relational dependencies between lsc-problems.

If we add, for example, the assumption that for the ability of the first and second subprocess of the algorithm to detect the relation S between two letters x_i and x_j the number of letters between x_i and x_j plays no role (formal $f_S, s_S \in \{0, k\}$) the surmise relations $\sqsubseteq_I'$ and $\sqsubseteq_{II}'$ derived from the model will become much stricter.

The surmise relation $\sqsubseteq_I'$ contains under this additional assumption the additional dependencies $3 \sqsubseteq_I' 1, 2, 4, 5, 6, 8, 9, 10, 11, 13, 14, 15, 17, 20$. None of these new relational dependencies is violated empirically in Experiment I. The surmise relation $\sqsubseteq_{II}'$ contains under this additional assumption, the additional relational dependencies $1 \sqsubseteq_{II}' 3, 5, 10, 12, 14, 15, 16, 22$. These new relational dependencies are violated empirically at most once in Experiment II. The observed symmetric distances of $\mathcal{W}_{\sqsubseteq_I'}$ and $\mathcal{W}_{\sqsubseteq_{II}'}$ to the data patterns in our experiments are identical with the observed symmetric distances of $\mathcal{W}_{\sqsubseteq_I}$ and $\mathcal{W}_{\sqsubseteq_{II}}$ to these data patterns, whereas $\mathcal{W}_{\sqsubseteq_I'}$ contains only 216 knowledge states and $\mathcal{W}_{\sqsubseteq_{II}'}$ contains 2919 knowledge states. So, the additional restriction $f_S, s_S \in \{0, k\}$ seems to be well suited by the data and improves the predictions of the model concerning the difficulty of lsc-problems.

Our considerations lead to the following two observations. First, our model predicts not all relational dependencies that seem to be immanent in our data sets. As we have seen this may result from the fact that we have not formulated enough restrictions on the possible parameter combinations. Second, an analysis of the experimental data can be used to search for additional restrictions on possible parameter combinations. Such additional restrictions can be incorporated in the model and can improve the models predictions concerning the difficulty of lsc-problems. We have not enforced such an analysis in detail for our two data sets because the high average solution frequency in our two experiments indicates that many of the presented lsc-problems are too easy for a large majority of subjects participating in these experiments. Therefore, an eventually existing difference in the difficulty of two lsc-problems may not be observable in our data sets.

ACKNOWLEDGEMENTS

This article is based on the authors doctoral dissertation at the University of Heidelberg, supported by a fellowship of the state Baden–Württemberg. The reported experiments were supported by Grant Lu 385/1 of the Deutsche Forschungsgemeinschaft to J. Lukas and D. Albert. I am grateful to B. Hirsmüller and D. Klein for their support in conducting the experimental investigations.

REFERENCES

Doignon, J. P. (1994). Knowledge spaces and skill assignments. In: G. Fischer & D. Laming (Eds.), *Contributions to mathematical psychology, psychometrics, and methodology* (pp. 111–121). Berlin: Springer.

Doignon, J. P., & Falmagne, J. C. (1985). Spaces for the assessment of knowledge. *International Journal of Man–Machine Studies*, *23*, 175–196.

Doignon, J.-P., & Falmagne, J.-C. (1998). *Knowledge spaces.* Berlin: Springer.

Falmagne, J. C., & Doignon, J. C. (1988). A markovian procedure for assessing the state of a system. *Journal of Mathematical Psychology*, *32*, 232–258.

Klahr, K., & Wallace, J.G. (1970). The development of serial completion strategies: An information processing analysis. *British Journal of Psychology*, *61*, 243–257.

Kotovsky, K., & Simon, H.A. (1973). Empirical tests of a theory of human acquisition of concepts for sequential patterns. *Cognitive Psychology*, *4*, 399–424.

Schneider, W. (1990). MEL user's guide: Computer techniques for real time experimentation. *Pittsburgh: Psychology Software Tools.*

Schrepp, M. (1993). *Über die Beziehung zwischen kognitiven Prozessen und Wissensräumen beim Problemlösen.* [On the relation between cognitive processes and knowledge spaces in problem solving.] Unpublished doctoral dissertation, University of Heidelberg, Germany.

Simon, H. A., & Kotovsky, K. (1963). Human acquisition of concepts for sequential patterns. *Psychological Review*, *70*, 534–546.

III

APPLICATIONS

7

Organizing and Controlling Learning Processes Within Competence-Performance Structures

Klaus Korossy
Ruprecht-Karls-Universität Heidelberg

In Chapter 5, we developed the concepts of a diagnostic and a union-stable diagnostic. It was demonstrated how a union-stable diagnostic can be constructed, and in principle, validated by collecting data and checking the observed problem-solving patterns. In the present chapter, further considerations of a more theoretical nature are presented on how a union-stable diagnostic can be used for organizing and controlling learning processes. Two main problems will be dealt with: First, the fundamental problem is analyzed, whether and how goal-directed adaptive instruction can be planned under conditions of non-unique competence assessment. Second, returning to the basic idea of a learning path, which is conceived as a chain of states, theoretical results on the representability of a (maximal) competence learning path by a (maximal) performance learning path are derived.

INTRODUCTION

In the context of learning and instruction, knowledge assessment is primarily applied for testing whether a certain teaching goal is achieved and for diagnosing the learner's current knowledge state (often called *student modeling*). Depending on the results of that diagnosis, further steps of instruction are selected. Optimally, the two stages of knowledge assessment and instruction are in tune with one another by being embedded in a sensitive feedback process and, thus, together enable a flexible, learner-centered application of a curriculum.

For the purpose of organizing teaching/learning processes, Doignon and Falmagne's knowledge structure theory seems to be inapplicable because of its purely behavioral character.[1] Because the knowledge structure theory does not relate solution dependencies between problems to an underlying theoretical concept of "knowledge" or "ability", observed differences in problem solving behavior can neither be theoretically explained nor prognostically used for predicting future solving behavior on another set of problems; moreover, to initiate a particular learning step, so that a previously unsolvable problem becomes solvable, the theory cannot give any guidance on which "piece of knowledge" has to be taught.

In Chapter 5, it was outlined how the knowledge structure theory may be integrated into traditional diagnostical approaches by extending it to a *competence-performance conception*[2] that can account for a more elaborated "cognitive" diagnosing. In that conception, a student's "knowledge state" in a particular domain of knowledge is modeled twofold: First, as a subset of problems the student is capable of solving, called the *performance state* of the student (comparable to the notion of a "knowledge state" in Doignon and Falmagne's theory); second, the student's *competence state* that comprises the nondirectly observable "competence", "ability" or "skills" underlying the student's observable problem solving behavior. If the competence states are presupposed to be constituted as subsets of a set of *elementary competencies*, then the family of competence states is called a *competence structure.* Accordingly, the family of all (empirically expectable) performance states is called a *performance structure.* The connection between the two structures is established through the basic concept of a *diagnostic* that links a competence structure and a performance structure by two mappings called *interpretation function* and *representation function*; the most elaborated concept is that of a *union-stable diagnostic* in which a *competence space* (that is a union-stable competence structure containing $\emptyset$ and the set of all elementary competencies as states) is related to a structurally analogous *performance space* via a union-preserving (not necessarily bijective) representation function.

In the current chapter, it is argued that the extended theoretical framework also provides a contribution for the task of organizing learning-goal oriented adaptive instruction. To describe the basic idea in the context of learning and instruction, each competence state as well as each performance state may be interpreted not only as a possible model for a student's (incomplete) "knowledge", the top state

[1] For an overview of this theoretical framework see Doignon and Falmagne (1998) or Falmagne, Koppen, Villano, Doignon, and Johannesen (1990).

[2] An elaboration of the competence-performance conception is found in Korossy (1993). Also, all relevant proofs omitted in the present chapter are carried out there.

representing the full knowledge/ability of an expert[3], but moreover may be considered as a subset of elementary *learning goals*. Under that teaching learning view, a transition from one state to another superordinate one may be conceived as a learning step. Thus the competence structure can be regarded as a goal-lattice or a family of sequences of learning states that can guide the instructional procedure in the sense of a '"curriculum". The performance structure can be regarded as the testing structure where success and failure of learning are observed.

Of course, a trouble-free learning teaching process requires at least that the underlying competence structure be didactically and psychologically validated, and the link between the curriculum and the performance level should be made precisely explicit. In Chapter 5, it was demonstrated how the task of validation can be accomplished on the basis of suitable representation concepts (the concepts of a diagnostic and a union-stable diagnostic). However, under the perspective of learning and instruction, some crucial problems concerning the connection between the competence and the performance structure remain to be analyzed. Two main problems are discussed here:

(1) The notion of "representation", expressed by the fundamental concept of a diagnostic, involves that observing a particular performance state does not necessarily enable the identification of a unique underlying competence state. In view of the organizing goal-oriented instruction procedures, it is necessary to investigate how this problem of "fuzzy competence assessment" can be managed.

(2) Within the framework of the competence-performance approach, goal directed knowledge instruction means that units of knowledge from along a sequence of competence states are imparted adaptively and the incremental progress of learning is evaluated on the performance level. The central question to be investigated here is whether and how an instructional plan on the level of competence (a path of competence states) corresponds to a series of increasing solution successes on problems (a path of performance states) and vice versa.

In the present chapter, the concept of central interest is that of a union-stable diagnostic. We will see that this concept may provide an appropriate basic structure for the objectives of goal-oriented adaptive knowledge instruction.

The following section recapitulates briefly several concepts and results from Chapter 5, in particular the central concept of a union-stable diagnostic. In the following sections, Problem (1) presented above is analyzed. In the final section, Problem (2) is discussed.

[3] This interpretation corresponds to the *overlay model* where a student's knowledge is conceived as part of an expert's knowledge.

THEORETICAL BACKGROUND

In this section, some basic concepts and results from the competence-performance view of the knowledge structures theory developed in Korossy (1993; see also Chapter 5, this volume) are briefly reviewed in order to facilitate reading. Let us first recall two pairs of concepts that are motivated by the interpretational background, but are at this point not yet structurally different.

- A pair $(\mathcal{E}, \mathcal{K})$ consisting of a non-empty, finite set $\mathcal{E}$ whose elements are called *elementary competencies*, and a family $\mathcal{K}$ of subsets of $\mathcal{E}$ called *competence states*, where $\cup\mathcal{K} = \mathcal{E}$, is said to be a *competence structure*.

  If $\emptyset, \mathcal{E} \in \mathcal{K}$ and $\mathcal{K}$ is stable under union, then $(\mathcal{E}, \mathcal{K})$ is called a *competence space*. Often $(\mathcal{E}, \mathcal{K})$ is denoted by $(\mathcal{K}, \cup)$.
- A pair $(\mathcal{A}, \mathcal{P})$ consisting of a non-empty, finite set $\mathcal{A}$ of *problems* and a family $\mathcal{P}$ of subsets of $\mathcal{A}$, called *performance states*, where $\cup\mathcal{P} = \mathcal{A}$, is said to be a *performance structure*.

  If $\emptyset, \mathcal{A} \in \mathcal{P}$ and $\mathcal{P}$ is stable under union, then $(\mathcal{A}, \mathcal{P})$ is called a *performance space*. The short denotation is $(\mathcal{P}, \cup)$.

These two concepts receive formally different positions within the concept of a *diagnostic* that involves the fundamental representation concept of the competence-performance approach.

DEFINITION 7.1 Let $\mathcal{K}$ be a set of competence states and $(\mathcal{A}, \mathcal{P})$ a performance structure. Furthermore, let

$k : \mathcal{A} \longrightarrow \wp(\mathcal{K})$ be a function that assigns to each problem $x \in \mathcal{A}$ a subset $k_x := k(x) \subseteq \mathcal{K}$ of competence states (those in each of which x is solvable) so that $k_x \neq \emptyset$ and $k_x \neq \mathcal{K}$,

$p : \mathcal{K} \longrightarrow \wp(\mathcal{A})$, with $p(\kappa) := \{x \in \mathcal{A} \mid \kappa \in k_x\}$ for each $\kappa \in \mathcal{K}$, be the uniquely determined function which assigns to each competence state $\kappa \in \mathcal{K}$ the set of all problems solvable in κ.

Assume now that the function p is surjective from $\mathcal{K}$ onto $p(\mathcal{K}) = \mathcal{P}$. Then $(\mathcal{A}, \mathcal{P})$ is called a *representation of* $\mathcal{K}$ (under p). The 5–tupel $(\mathcal{K}, \mathcal{A}, \mathcal{P}, k, p)$ is said to be a *diagnostic*. The function k is called *interpretation function for* $(\mathcal{A}, \mathcal{P})$, the fnction p is called *representation function for* $\mathcal{K}$. If the competence states of $\mathcal{K}$ are subsets of a set $\mathcal{E}$ of elementary competencies (that is if $(\mathcal{E}, \mathcal{K})$ is a competence structure), then the diagnostic is written as a 6–tupel $(\mathcal{E}, \mathcal{K}, \mathcal{A}, \mathcal{P}, k, p)$.

The representational concept of central interest in the present chapter is that of a *union-stable diagnostic*.

DEFINITION 7.2 Let $(\mathcal{E}, \mathcal{K}, \mathcal{A}, \mathcal{P}, k, p)$ be a diagnostic, $(\mathcal{K}, \cup)$ a competence space, $(\mathcal{P}, \cup)$ a performance space, and $p : \mathcal{K} \longrightarrow \mathcal{P}$ a union-preserving representation function, i.e.

$$p(\kappa \cup \lambda) = p(\kappa) \cup p(\lambda), \qquad \textit{for all } \kappa, \lambda \in \mathcal{K}.$$

Then the representation $(\mathcal{P}, \cup)$ of $(\mathcal{K}, \cup)$ is called a *union-preserving representation* of $(\mathcal{K}, \cup)$; the diagnostic $(\mathcal{E}, \mathcal{K}, \mathcal{A}, \mathcal{P}, k, p)$ is called a *union-stable diagnostic.*

A *learning step* can be conceived as a transition from one competence state (performance state, respectively) to a "superordinate state" that includes more competencies (solvable problems, respectively) than the previous one. Therefore, in some sections of this chapter, *order structures* on the competence space and on the performance space will become important. Let us remark here that in each semi-lattice $(\mathcal{K}, \cup)$ the subset relation $\subseteq$ can be defined by

$$\kappa \subseteq \lambda :\Longleftrightarrow \kappa \cup \lambda = \lambda, \qquad \textit{for all } \kappa, \lambda \in \mathcal{K}. \tag{1}$$

We note for later use:

LEMMA 7.1 Let $(\mathcal{E}, \mathcal{K}, \mathcal{A}, \mathcal{P}, k, p)$ be a union-stable diagnostic, and consider the competence space $(\mathcal{K}, \cup)$ and the performance space $(\mathcal{P}, \cup)$ as partially ordered sets $(\mathcal{K}, \subseteq)$ and $(\mathcal{P}, \subseteq)$, with the subset relation $\subseteq$ as the partial order introduced according to (1). Then the union-preserving representation function p is especially order-preserving, that is

$$\kappa \subseteq \lambda \Longrightarrow p(\kappa) \subseteq p(\lambda), \qquad \textit{for all } \kappa, \lambda \in \mathcal{K}.$$

In the subsequent considerations we refer to a simple property of *union-stable problems* that guarantee a union-preserving representation of a competence space $(\mathcal{K}, \cup)$.

PROPOSITION 7.1 Let $(\mathcal{K}, \cup)$ be a competence space with the set $\mathcal{E}$ of elementary competencies. A problem x with interpretation $k_x \subseteq \mathcal{K}$ is a union-stable problem if and only if there exists a non-empty subset $\varphi_x \subseteq \mathcal{E}$ of elementary competencies, so that

$$k_x = \{\nu \in \mathcal{K} \mid \nu \cap \varphi_x \neq \emptyset\}.$$

Such a (not necessarily unique) set φ_x ocurring in the above proposition is called a *generating set* for k_x.

In the following three sections, we turn to the problem of "fuzzy competence assessment". Then, the problems concerning the correspondence between learning paths on the competence and those on the performance level is the topic of discussion.

THE PROBLEM OF FUZZY COMPETENCE ASSESSMENT

In the basic concept of a diagnostic, the following problem is inherent: The representation of a competence structure through a performance structure is conceived as a unique but not necessarily one-to-one mapping. Thus, a competence modeling is allowed to be represented by a coarser-grained image on the performance level. However, in view of the task of assessing the knowledge state of a person, this "liberal" representation concept involves that observing a performance state often yields no more than an imprecise ("fuzzy") diagnosis of the underlying competence states. In this and the subsequent two sections, we will show that a *union-stable diagnostic* in spite of this fuzziness is an appropriate framework for organizing goal-directed controlled knowledge instruction.

Partitions on the set of competence states

First, we focus our attention on *relations* and *partitions* that are—as a consequence of the weak representation concept—generated on a given set of competence states by representing performance structures. A main role is played by the equivalence relation *kernel of* p denoted by $Ker(p)$ that is induced by the respective representation function p.

Let $(\mathcal{K}, \mathcal{A}, \mathcal{P}, k, p)$ be a diagnostic. We consider the binary relation $Ker(p)$ on $\mathcal{K}$ induced by the representation function $p : \mathcal{K} \longrightarrow \mathcal{P}$. The usual definition of $Ker(p)$,

$$(\kappa, \lambda) \in Ker(p) \; :\Longleftrightarrow \; p(\kappa) = p(\lambda) , \qquad \textit{for all } \kappa, \lambda \in \mathcal{K},$$

immediately reveals that $Ker(p)$ is reflexive, symmetric and transitive, hence an equivalence relation on $\mathcal{K}$. In general, each equivalence relation ϕ on $\mathcal{K}$ gives rise to a partition of the set $\mathcal{K}$ into a set of pairwise disjoint equivalence classes of the form

$$[\kappa]_\phi := \{\lambda \in \mathcal{K} \mid (\kappa, \lambda) \in \phi\} , \qquad \textit{for all } \kappa \in \mathcal{K},$$

that in our context are called *competence classes*; the *partition* (*quotient set of* $\mathcal{K}$ *modulo* ϕ)

$$\mathcal{K}/\phi := \{[\kappa]_\phi \mid \kappa \in \mathcal{K}\}$$

is called *competence partition*; for the surjective mapping

$$q : \begin{cases} \mathcal{K} \longrightarrow \mathcal{K}/\phi \\ \kappa \longmapsto q(\kappa) := [\kappa]_\phi \end{cases} , \qquad (2)$$

that assigns to each competence state its competence class in $\mathcal{K}/\phi$, we adopt the usual expression *quotient map*. The relation *kernel of* q, abbreviated by $Ker(q)$, is also an equivalence relation on $\mathcal{K}$, and it should be noted that

$$(\kappa, \lambda) \in \phi \iff [\kappa]_\phi = [\lambda]_\phi \iff (\kappa, \lambda) \in Ker(q) .$$

A well-known theorem from universal algebra then specializes to the following proposition.

PROPOSITION 7.2 Let $(\mathcal{K}, \mathcal{A}, \mathcal{P}, k, p)$ be a diagnostic; for $\phi := Ker\,(p)$ let $q : \mathcal{K} \longrightarrow \mathcal{K}/\phi$ be the quotient map according to (2). Then, the map $h : \mathcal{K}/\phi \longrightarrow \mathcal{P}$, given by

$$h([\kappa]_\phi) := p(\kappa)\,, \qquad \textit{for all } \kappa \in \mathcal{K}\,,$$

is well-defined, and it holds $p = h \circ q$, that is, the diagram

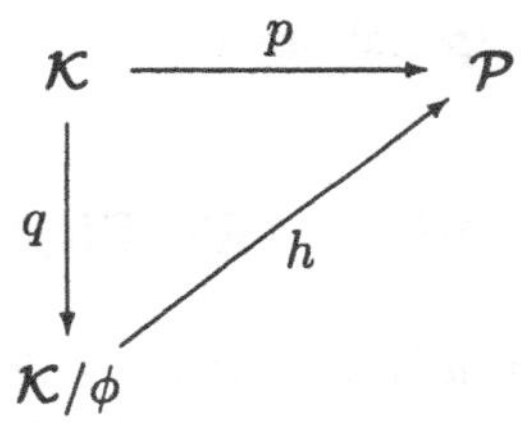

commutes. Furthermore, the map h is bijective and $\phi = Ker\,(q)$.

PROOF: The map h is well-defined: Let be $\kappa, \lambda \in \mathcal{K}$ with $[\kappa]_\phi = [\lambda]_\phi$, thus $p(\kappa) = p(\lambda)$. Then, $h([\kappa]_\phi) = p(\kappa) = p(\lambda) = h([\lambda]_\phi)$. Clearly, $p = h \circ q$ by definition. The map h is *surjective* because p is surjective; since

$$h([\kappa]_\phi) = h([\lambda]_\phi) \implies p(\kappa) = p(\lambda) \implies (\kappa, \lambda) \in \phi \implies [\kappa]_\phi = [\lambda]_\phi\,,$$

h is also *injective*. □

Proposition 7.2 characterizes the connection between the family of competence states and the family of performance states. As a consequence of the not necessarily bijective representation function p, the set $\mathcal{K}$ is divided into competence classes, each of which contains competence states that cannot be distinguished by means of the performance states in $\mathcal{P}$. Looking at the inverse map h^- of $h : \mathcal{K}/\phi \longrightarrow \mathcal{P}$ from Proposition 7.2, defined by

$$h^- : \begin{cases} \mathcal{P} \longrightarrow \mathcal{K}/\phi \\ Z \longmapsto h^-(Z) := \{\kappa \in \mathcal{K} \mid p(\kappa) = Z\} \end{cases}, \tag{3}$$

we see the following: For each performance state $Z \in \mathcal{P}$, the competence class $h^-(Z)$ contains all those competence states that "generate" or "explain" Z, that is to say, those that can be "diagnosed" as competence states possibly underlying the observed performance state of a person.

As just seen above, the partition $\mathcal{K}/\phi$ is to be considered as the actual goal structure of the competence diagnosis. The grade of "graininess" of the partition of $\mathcal{K}$ induced by the representing performance structure determines how precise a diagnosis will be in the best case. In domain-specific applications of the competence-performance approach, it has to be weighed up under conditions of economy whether a realized grade of graininess is sufficient in view of the intended purposes. If necessary, the graininess may be refined through refining the performance representation by means of extending the set of problems—in the extreme case up to the uniqueness of competence diagnoses.

The preceding considerations provide no more than an ordering schema for "fuzzy competence diagnoses". In the next two sections, we continue dealing with competence partitions for the special case of union-stable diagnostics.

CONGRUENCE PARTITIONS

Let $(\mathcal{E}, \mathcal{K}, \mathcal{A}, \mathcal{P}, k, p)$ be a *union-stable diagnostic*, with the performance space $(\mathcal{P}, \cup)$ being a union-preserving as well as an order-preserving (with respect to the partial order $\subseteq$) representation of the competence space $(\mathcal{K}, \cup)$. Furthermore, let $\mathcal{K}/\mathit{Ker}(p)$ be the competence partition induced by the representation function p. In the following, we show how and in what sense the partition $\mathcal{K}/\mathit{Ker}(p)$, through defining an operation of union and a partial order, may be conceived as a "structure-preserving coarsement" of the competence space $(\mathcal{K}, \cup)$. From this will follow several important consequences that reveal how in applications of union-stable diagnostics the problem of fuzzy diagnoses can be overcome.

Congruence relations on the competence space

Let us start with establishing that, if $(\mathcal{E}, \mathcal{K}, \mathcal{A}, \mathcal{P}, k, p)$ is a union-stable diagnostic, then the equivalence relation $\mathit{Ker}(p)$ is in a definite sense compatible with the operation $\cup$ on $\mathcal{K}$.

LEMMA 7.2 Let $(\mathcal{E}, \mathcal{K}, \mathcal{A}, \mathcal{P}, k, p)$ be a union-stable diagnostic. Then, the equivalence relation $\tau := \mathit{Ker}(p)$ is compatible with the operation $\cup$ on $\mathcal{K}$, that is, for κ, κ', λ, $\lambda' \in \mathcal{K}$ holds:

$$(\kappa, \kappa') \in \tau \wedge (\lambda, \lambda') \in \tau \implies (\kappa \cup \lambda, \kappa' \cup \lambda') \in \tau .$$

PROOF: Let κ, κ', λ, $\lambda' \in \mathcal{K}$, with (κ, κ'), $(\lambda, \lambda') \in \mathit{Ker}(p)$, i.e. $p(\kappa) = p(\kappa')$ and $p(\lambda) = p(\lambda')$. Using that p is union-preserving, we then get

$$p(\kappa \cup \lambda) = p(\kappa) \cup p(\lambda) = p(\kappa') \cup p(\lambda') = p(\kappa' \cup \lambda'),$$

hence $(\kappa \cup \lambda, \kappa' \cup \lambda') \in \mathit{Ker}(p)$. □

Adopting the usual mathematical terminology, we generally term an equivalence relation τ on a $\cup$-semilattice $(\mathcal{K}, \cup)$ as being compatible with $\cup$ in the sense of Lemma 7.2 a *congruence relation* (*relative to* $\cup$) *on* $\mathcal{K}$; the partition $\mathcal{K}/\tau$ generated by a congruence relation τ on $\mathcal{K}$ is called a *congruence partition*, its classes are called *congruence classes*.

EXAMPLE 7.1 Let $(\mathcal{K}, \cup)$ be the competence space with[4]

$$\mathcal{K} := \{\emptyset, 1, 3, 12, 13, 15, 23, 45, 123, 125, 135, 145, 235, 345, \\ 1234, 1235, 1245, 1345, 2345, 12345\}.$$

On $\mathcal{K}$ let $\mathcal{K}/\tau$ be the following competence partition modulo an equivalence relation τ:

$$\mathcal{K}/\tau := \{\{\emptyset\}, \{1\}, \{3\}, \{12\}, \{15\}, \{13, 23, 123\}, \{125\}, \{45\}, \\ \{135, 235, 1235\}, \{145\}, \{345\}, \{1234, 1245, 1345, 2345, 12345\}\}.$$

The partition is a *congruence partition* according to the definition. For instance, $13 \cup 45 = 1345$ is congruent to $23 \cup 45 = 2345$. ♣

A union operation on the congruence partition

Now, let $(\mathcal{K}, \cup)$ be a $\cup$-semilattice, τ a congruence relation (relative to $\cup$) on $\mathcal{K}$, $\mathcal{K}/\tau$ the associated congruence partition on $\mathcal{K}$. Then we define an operation $\sqcup$ on $\mathcal{K}/\tau$ for every two congruence classes $[\kappa]_\tau$, $[\lambda]_\tau \in \mathcal{K}/\tau$ by

$$[\kappa]_\tau \sqcup [\lambda]_\tau := [\kappa \cup \lambda]_\tau . \tag{4}$$

The operation $\sqcup$ is well-defined when the definition is independent of the elements chosen to represent the congruence classes, that is, when for all κ, κ', λ, $\lambda' \in \mathcal{K}$ holds:

$$[\kappa]_\tau = [\kappa']_\tau \wedge [\lambda]_\tau = [\lambda']_\tau \implies [\kappa \cup \lambda]_\tau = [\kappa' \cup \lambda']_\tau .$$

This, however, follows immediately from the defining property of the congruence relation τ in Lemma 7.2. The defining property then also implies that the quotient map $q : \mathcal{K} \longrightarrow \mathcal{K}/\tau$ is *union-preserving*, that is,

$$q(\kappa \cup \lambda) \equiv [\kappa \cup \lambda]_\tau = [\kappa]_\tau \sqcup [\lambda]_\tau \equiv q(\kappa) \sqcup q(\lambda), \qquad \textit{for all } \kappa, \lambda \in \mathcal{K}.$$

As an image of the semilattice $(\mathcal{K}, \cup)$ under the union-preserving function q, apparently $\mathcal{K}/\tau$ is a $\sqcup$-semilattice, that is, $\mathcal{K}/\tau$ is *stable* under $\sqcup$ and, according

[4] For a short notation, the competence states—as well as the performance states later on—are compactly denoted as sequence of symbols instead of as sets.

to the definition in (4) and the properties of $\cup$ on $\mathcal{K}$, $\sqcup$ is *commutative*, *associative* and *idempotent*; if $\mathcal{E}$ is the top element in $\mathcal{K}$, then $[\mathcal{E}]_\tau$ is the top element in $\mathcal{K}/\tau$; if $\mathcal{K}$ has a bottom element $\emptyset$ (as it is required when $\mathcal{K}$ is to be a competence space), then $[\emptyset]_\tau$ is the bottom element in $\mathcal{K}/\tau$. Finally, each congruence class itself is closed under $\cup$, hence (as a subset of $\mathcal{K}$) a semilattice.

We summarize these considerations in the next proposition.

PROPOSITION 7.3 Let $(\mathcal{K}, \cup)$ be a semilattice with bottom element $\emptyset$ and top element $\mathcal{E}$, τ a congruence relation relative to $\cup$ on $\mathcal{K}$, $\mathcal{K}/\tau$ the associated congruence partition, and $\sqcup$ the union operation on $\mathcal{K}/\tau$ defined by (4) Then the following statements hold:

(1) $(\mathcal{K}/\tau, \sqcup)$ is a semilattice with top element $[\mathcal{E}]_\tau$ and bottom element $[\emptyset]_\tau$.
(2) Each congruence class $[\kappa]_\tau \in \mathcal{K}/\tau$ is a semilattice relative to $\cup$.
(3) The quotient map $q : \mathcal{K} \longrightarrow \mathcal{K}/\tau$ is union-preserving, that is,
$q(\kappa \cup \lambda) = q(\kappa) \sqcup q(\lambda)$, *for all* $\kappa, \lambda \in \mathcal{K}$.

EXAMPLE 7.2 With the operation $\sqcup$ according to (4), the congruence partition $\mathcal{K}/\tau$ from Example 7.1 constitutes the semilattice $(\mathcal{K}/\tau, \sqcup)$ with bottom element $[\emptyset]_\tau$ and top element

$$[12345]_\tau = \{1234, 1245, 1345, 2345, 12345\}.$$

That the operation $\sqcup$ is well-defined (hence the quotient map q union-preserving), is seen for example with the congruence classes $\{13, 23, 123\}$ and $\{45\}$:

$$\begin{aligned}
[13]_\tau \sqcup [45]_\tau &= [13 \cup 45]_\tau = [1345]_\tau, \\
[23]_\tau \sqcup [45]_\tau &= [23 \cup 45]_\tau = [2345]_\tau, \\
[123]_\tau \sqcup [45]_\tau &= [123 \cup 45]_\tau = [12345]_\tau;
\end{aligned}$$

the results are all identical, independent of the chosen elements of the equivalence classes, namely

$$[1345]_\tau = [2345]_\tau = [12345]_\tau = \{1234, 1245, 1345, 2345, 12345\}.$$

Finally, one easily verifies that each congruence class itself constitutes a semilattice under $\cup$. ♣

Following Proposition 7.3, a competence space $(\mathcal{K}, \cup)$ is partitioned by a congruence relation (an equivalence relation τ compatible with $\cup$) in just such a way that the competence partition $\mathcal{K}/\tau$, equipped with an appropriately defined operation $\sqcup$, appears as a coarser-grained image of the original competence space. The next subsection will clarify whether an analagous result also holds under ordering perspectives.

Congruence partitions as partially ordered sets

As we know from (1), in any semilattice $(\mathcal{K}, \cup)$ the partial order $\subseteq$ may be defined by

$$\kappa \subseteq \lambda :\Longleftrightarrow \kappa \cup \lambda = \lambda\,, \qquad \textit{for all } \kappa,\, \lambda \in \mathcal{K}\,.$$

Thus, a competence space $(\mathcal{K}, \cup)$ can always be considered as a partially ordered set $(\mathcal{K}, \subseteq)$ with greatest element $\mathcal{E}$ and least element $\emptyset$. Now, if $\mathcal{K}/\tau$ is a congruence partition, then the question arises how the ordering structure might be transferred from $\mathcal{K}$ to the partition $\mathcal{K}/\tau$.

Apparently, also on the semilattice $(\mathcal{K}/\tau, \sqcup)$ a binary relation $\sqsubseteq$ may be obtained formally by the same way as $\subseteq$ was defined on $(\mathcal{K}, \cup)$. For $[\kappa]$, $[\lambda] \in \mathcal{K}/\tau$ let

$$[\kappa] \sqsubseteq [\lambda] \;:\Longleftrightarrow\; [\kappa] \sqcup [\lambda] = [\lambda]\,. \tag{5}$$

It is easy to prove that $\sqsubseteq$ is reflexive, antisymmetric, and transitive, hence a partial order on $\mathcal{K}/\tau$; additionally, the quotient map $q : \mathcal{K} \longrightarrow \mathcal{K}/\tau$ is order-preserving. Moreover, the fact that each congruence class $[\omega]_\tau \in \mathcal{K}/\tau$ is convex relative to $\subseteq$ will gain in importance, that is, for κ, $\lambda \in [\omega]_\tau$ and any $\mu \in \mathcal{K}$ with $\kappa \subseteq \mu \subseteq \lambda$ it follows that $\mu \in [\omega]_\tau$. To sum up, we formulate the following proposition.

PROPOSITION 7.4 Let $(\mathcal{K}, \cup)$ be a semilattice with bottom element $\emptyset$ and top element $\mathcal{E}$ and with the partial order $\subseteq$; further, let $\mathcal{K}/\tau$ be a congruence partition on $(\mathcal{K}, \cup)$ with the union operation $\sqcup$ and the relation $\sqsubseteq$ defined according to (5). Then the following hold:

(1) $(\mathcal{K}/\tau, \sqsubseteq)$ is a partially ordered set with the greatest element $[\mathcal{E}]_\tau$ and the least element $[\emptyset]_\tau$.
(2) Each congruence class $[\omega] \in \mathcal{K}/\tau$ is convex relative to $\subseteq$, that is, for all $\kappa, \lambda \in [\omega]_\tau$ and $\mu \in \mathcal{K}$ holds: $\kappa \subseteq \mu \subseteq \lambda \implies \mu \in [\omega]_\tau$.
(3) The quotient map $q : \mathcal{K} \longrightarrow \mathcal{K}/\tau$ is order-preserving, that is,
$\kappa \subseteq \lambda \implies q(\kappa) = [\kappa]_\tau \sqsubseteq [\lambda]_\tau = q(\lambda)\,, \qquad \textit{for all } \kappa,\, \lambda \in \mathcal{K}\,.$

Obviously, Proposition 7.4 parallels Proposition 7.3. Therefore, the partially ordered partition $(\mathcal{K}/\tau, \sqsubseteq)$ may be regarded as a coarser-grained image of the partially ordered set $(\mathcal{K}, \subseteq)$ as well.

In the context of learning processes, transitions from a lower to a higher competence class are of central importance. Those transitions can be described with the aid of the above introduced relation $\sqsubseteq$ on $\mathcal{K}/\tau$. The following proposition reveals how this relation is to be handled with respect to $\subseteq$.[5]

[5] The somewhat technical proof is omitted here.

PROPOSITION 7.5 Let $(\mathcal{K}, \cup)$ be a semilattice with the partial order $\subseteq$, $\mathcal{K}/\tau$ a congruence partition with the operation $\sqcup$ and the partial order $\sqsubseteq$ according to (5). For all $[\kappa]_\tau$, $[\lambda]_\tau \in \mathcal{K}/\tau$ the following are equivalent:

(1) $[\kappa]_\tau \sqsubseteq [\lambda]_\tau$;
(2) $\exists \kappa' \in [\kappa]_\tau \, \exists \lambda' \in [\lambda]_\tau \, (\kappa' \subseteq \lambda')$;
(3) $\forall \kappa' \in [\kappa]_\tau \, \exists \lambda' \in [\lambda]_\tau \, (\kappa' \subseteq \lambda')$;
(4) $\max [\kappa]_\tau \subseteq \max [\lambda]_\tau$.[6]

EXAMPLE 7.3 Consider the congruence partition $\mathcal{K}/\tau$ from Examples 7.1 and 7.2. The statements of Proposition 7.5 on the partial order $\sqsubseteq$ on a congruence partition are verified for example by the congruence classes $[13]_\tau = \{13, 23, 123\}$ and $[135]_\tau = \{135, 235, 1235\}$ of the congruence partition $\mathcal{K}/\tau$. Note that it is incorrect that $\forall \kappa' \in [\kappa] \, \forall \lambda' \in [\lambda] \, (\kappa' \subseteq \lambda')$. ♣

The results of our considerations up to this point now suggest some interesting conclusions. When the partitioning of a competence space $(\mathcal{K}, \cup)$ originated by a congruence relation τ is interpreted as an expression of the "fuzziness of competence diagnoses", then the properties of the congruence partition $\mathcal{K}/\tau$ have several important consequences for competence-based adaptive instruction:

- If a person's competence state κ is not exactly identified but diagnosed as being an element of a particular congruence class $[\kappa]_\tau$, then, when teaching some elementary competency ε, it can uniquely be predicted in what congruence class the resulting state of the person will be located.
- If for a person, whose not-exactly identified competence state is an element of a particular competence class $[\kappa]_\tau$, a learning step to a target class $[\lambda]_\tau$ with $[\kappa]_\tau \sqsubseteq [\lambda]_\tau$, is to be induced, then the uniquely determined elementary competencies from the set $\max [\lambda]_\tau \setminus \max [\kappa]_\tau$ are to be taught.

Let it be emphasized that these statements in general do not apply to non-union-stable diagnostics. Counterexamples are easily constructed.

The results for congruence partitions on competence spaces particularly characterize, for a union-stable diagnostic $(\mathcal{E}, \mathcal{K}, \mathcal{A}, \mathcal{P}, k, p)$, those congruence partitions that are generated by the congruence relation $Ker(p)$ on $\mathcal{K}$. The next section directs attention to the link between the congruence partition $\mathcal{K}/Ker(p)$ and the performance space $\mathcal{P}$.

[6] $\max [\kappa]$ in this context is the greatest element of the equivalence class $[\kappa]$ relative to set inclusion $\subseteq$.

THE FUNDAMENTAL THEOREM FOR UNION-STABLE DIAGNOSTICS

Let $(\mathcal{E}, \mathcal{K}, \mathcal{A}, \mathcal{P}, k, p)$ be a union-stable diagnostic with the competence space $(\mathcal{K}, \cup)$ and the performance space $(\mathcal{P}, \cup)$, both partially ordered by $\subseteq$; let $\mathcal{K}/\,Ker\,(p)$ be the competence partition on $\mathcal{K}$ induced by the congruence relation $Ker\,(p)$. After characterizing specifically the partition $\mathcal{K}/\,Ker\,(p)$ by Propositions 7.3 and 7.4, we now turn to investigate the relation between the congruence partition $\mathcal{K}/\,Ker\,(p)$, structured by $\sqsubseteq$ and $\sqcup$, and the representing performance space $\mathcal{P}$.

For this purpose consider the diagram from Proposition 7.2

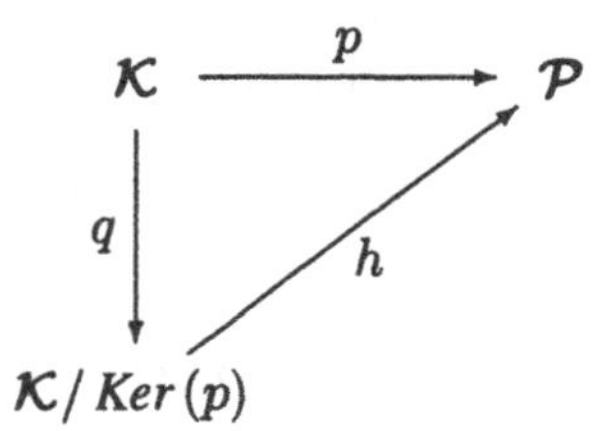

with the quotient map $q : \mathcal{K} \longrightarrow \mathcal{K}/\,Ker\,(p)$ from 2 and the map

$$h : \begin{cases} \mathcal{K}/\,Ker\,(p) \longrightarrow \mathcal{P} \\ [\kappa] \longmapsto h([\kappa]) := p(\kappa) \end{cases} . \tag{6}$$

According to Proposition 7.2, h is well-defined, and with the property $h \circ q = p$ of h the above diagram commutes; moreover, h is bijective. Completing Proposition 7.2 by statements on preserving-properties of h leads to the following theorem referred to as the *Fundamental Theorem for Union-Stable Diagnostics*.

PROPOSITION 7.6 Let $(\mathcal{E}, \mathcal{K}, \mathcal{A}, \mathcal{P}, k, p)$ be a union-stable diagnostic and $\mathcal{K}/\,Ker\,(p)$ the competence partition modulo $Ker\,(p)$, equipped with the operation $\sqcup$ and the partial order $\sqsubseteq$; let $q : \mathcal{K} \longrightarrow \mathcal{K}/\,Ker\,(p)$ be the associated quotient map and $h : \mathcal{K}/\,Ker\,(p) \longrightarrow \mathcal{P}$ the map well-defined via (6). Then, the following hold:

(1) The map h is a $\sqcup/\cup$ –isomorphism with

$$h([\kappa] \sqcup [\lambda]) = p(\kappa) \cup p(\lambda)\,, \qquad \textit{for all}\;\; [\kappa],\, [\lambda] \in \mathcal{K}/\,Ker\,(p)\,.$$

(2) The map h is a $\sqsubseteq / \subseteq$ –order-isomorphism with

$$[\kappa] \sqsubseteq [\lambda] \iff p(\kappa) \subseteq p(\lambda)\,, \qquad \textit{for all}\;\; [\kappa],\, [\lambda] \in \mathcal{K}/\,Ker\,(p)\,.$$

PROOF: (1) After Proposition 7.2, it only remains to prove that h is union-preserving. Because p is union-preserving and q according to Proposition 7.3 (3) is also union-preserving, we have for $[\kappa], [\lambda] \in \mathcal{K}/Ker(p)$

$$h([\kappa] \sqcup [\lambda]) \equiv h(q(\kappa) \sqcup q(\lambda)) = h(q(\kappa \cup \lambda)) = p(\kappa \cup \lambda) = p(\kappa) \cup p(\lambda).$$

(2) For $\kappa, \lambda \in \mathcal{K}$, $[\kappa], [\lambda] \in \mathcal{K}/Ker(p)$ is

$$\begin{aligned} [\kappa] \sqsubseteq [\lambda] \quad &:\Longleftrightarrow \quad [\kappa] \sqcup [\lambda] = [\lambda] \iff [\kappa \cup \lambda] = [\lambda] \\ &\iff \quad p(\kappa) \cup p(\lambda) = p(\kappa \cup \lambda) = p(\lambda) \\ &\iff \quad p(\kappa) \subseteq p(\lambda). \end{aligned}$$

□

Considering the *strict partial order* $\sqsubset$ associated with $\sqsubseteq$, one immediately obtains from Proposition 7.6 (2) the following corollary.

COROLLARY 7.1 Under the conditions of Proposition 7.6, for all $\kappa, \lambda \in \mathcal{K}$, $[\kappa], [\lambda] \in \mathcal{K}/Ker(p)$

$$[\kappa] \sqsubset [\lambda] \iff h([\kappa]) = p(\kappa) \subset p(\lambda) = h([\lambda]).$$

To illustrate the statements on congruence partitions on a competence space and the statements of the Fundamental Theorem for Union-Stable Diagnostics, the following detailed example seems appropriate.

EXAMPLE 7.4 The competence space $(\mathcal{K}, \cup)$ from Examples 7.1 and 7.2 now is extended into a union-stable diagnostic $(\mathcal{E}, \mathcal{K}, \mathcal{A}, \mathcal{P}, k, p)$ through interpreting a set $\mathcal{A}$ of problems in $\mathcal{K}$. The components of the diagnostic $(\mathcal{E}, \mathcal{K}, \mathcal{A}, \mathcal{P}, k, p)$ are shown in Table 1 as an overview. The top section of Table 1 presents the competence space $(\mathcal{E}, \mathcal{K})$ with

$$\begin{aligned} &\mathcal{E} := \{1, 2, 3, 4, 5\} \text{ (the family of elementary competencies)} \\ &\mathcal{K} := \{\emptyset, 1, 3, 12, 13, 15, 23, 45, 123, 125, 135, 145, 235, 345, \\ &\qquad 1234, 1235, 1245, 1345, 2345, 12345\}. \end{aligned}$$

The middle section of Table 1 shows the interpretation of a set $\mathcal{A} = \{a, b, c, d, e\}$ of problems in $\mathcal{K}$. For each problem $x \in \mathcal{A}$ a generating set φ_x and the corresponding interpretation $k_x \subseteq \mathcal{K}$ are listed. This establishment of the interpretation $k : \mathcal{A} \longrightarrow \wp(\mathcal{K})$ assures according to Proposition 7.1 that the resulting diagnostic $(\mathcal{E}, \mathcal{K}, \mathcal{A}, \mathcal{P}, k, p)$ is in fact union-stable. The values of the representation function $p : \mathcal{K} \longrightarrow \mathcal{P}$ induced by k can be found in the last line of Table 1. Competence states that are represented by the same performance state together make up a congruence class and are ordered in columns in the top section of Table 1. The family

$$\mathcal{P} := \{\emptyset, a, ab, ae, bc, abc, abe, cde, abce, acde, bcde, abcde\}$$

TABLE 1
Example: The Union-Stable Diagnostic $(\mathcal{E}, \mathcal{K}, \mathcal{A}, \mathcal{P}, k, p)$

	$\kappa \in \mathcal{K}$	$\emptyset$	1	3	12	13	15	45	125	135	145	345	1234
						23				235			1245
						123				1235			1345
													2345
													12345
$x \in \mathcal{A}$	φ_x												
a	$\{1,2\}$		◇		◇	◇	◇		◇	◇	◇		◇
b	$\{2,3\}$			◇	◇	◇			◇	◇		◇	◇
c	$\{3,4\}$			◇		◇		◇		◇	◇	◇	◇
d	$\{4\}$							◇			◇	◇	◇
e	$\{4,5\}$						◇	◇	◇	◇	◇	◇	◇
	$p(\kappa)$	$\emptyset$	a	bc	ab	abc	ae	cde	abe	$abce$	$acde$	$bcde$	$abcde$

The top section of the table shows the family $\mathcal{K}$ of competence states. The middle section shows the interpretation of a set $\mathcal{A}$ of problems in $\mathcal{K}$. For each problem $x \in \mathcal{A}$, a generating set φ_x and the corresponding interpretation k_x in $\mathcal{K}$, symbolized by ◇, is listed. The values of the representation function p induced by k can be found in the last line of the table. Competence states that are represented by the same performance state together make up a congruence class and are ordered in columns in the top section of the table.

of resulting performance states constitutes a performance space $(\mathcal{P}, \cup)$ as a union-preserving representation of the competence space $(\mathcal{K}, \cup)$ under the representation function p.

Fig. 1 illustrates the theoretical results on congruence partitions and the statements from the Fundamental Theorem for Union-Stable Diagnostics. It shows the congruence partition modulo $Ker(p)$ on the competence space $(\mathcal{K}, \cup)$ as a Hasse diagram relative to $\sqsubseteq$, where above each congruence class the corresponding performance state is depicted. ♣

Interpretation of the Fundamental Theorem

The Fundamental Theorem for Union-Stable Diagnostics constitutes the formal basis for the practical application of union-stable diagnostics in the context of qualitative knowledge assessment and instruction. Under conditions of restricted

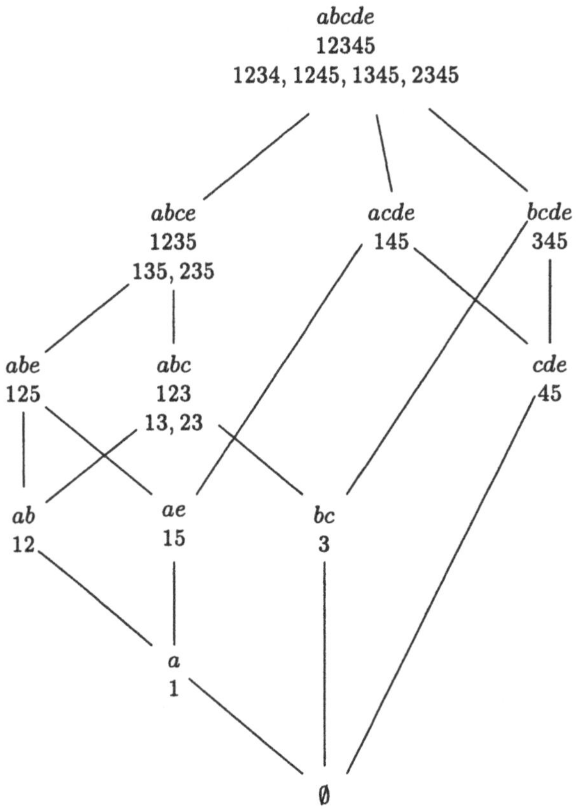

FIG. 1. Congruence partition $\mathcal{K}/\,Ker(p)$ of the union-stable diagnostic $(\mathcal{E}, \mathcal{K}, \mathcal{A}, \mathcal{P}, k, p)$ from the example, portrayed as a Hasse diagram relative to the partial order $\sqsubseteq$; above each congruence class the corresponding performance state is depicted.

performance observation, a given competence space is reorganized as a coarser-grained image of itself in such a way[7] that its essential structural properties are preserved and that the resulting competence partition is order-isomorphic and union-isomorphic to the representing performance space.

Extending from these fundamental properties, the following possibilities for practical use of union-stable diagnostics are opened up:

- As with each diagnostic, the "empirically evoked" competence partition (modulo the kernel of the representation function) can serve as an order-

[7] Technically as the congruence partition modulo the kernel of the representation function

ing schema for diagnosing the competence states possibly underlying an observed performance state.

- Additionally, in a union-stable diagnostic any surmise on a transition from a lower to a higher competence class can be formulated and tested as an empirical hypothesis on the transition between corresponding performance states.
- Moreover, because in a union-stable diagnostic even an order-isomorphism from the competence partition to the performance space exists, conversely, each performance transition corresponds to a transition between the associated competence classes.
- Additionally, because in a union-stable diagnostic the relation of the competence partition and the performance space is not only order- but also union-isomorphic, the preceding statement can be strengthened in the following sense: If a transition from a lower performance state to a higher performance state is to be induced (so that at least one more problem can be solved), then there exist uniquely determined elementary competencies that should bring about the intended competence transition.[8]

In the next section, the perspectives for competence-based instruction within the framework of union-stable diagnostics is pursued under the aspect of planning longer learning sequences.

PLANNING AND CONTROLLING LEARNING SEQUENCES

In this section, the competence space $(\mathcal{K}, \cup)$ and the performance space $(\mathcal{P}, \cup)$ in a union-stable diagnostic $(\mathcal{E}, \mathcal{K}, \mathcal{A}, \mathcal{P}, k, p)$ are conceived as partially ordered sets $(\mathcal{K}, \subseteq)$ and $(\mathcal{P}, \subseteq)$, and the representation function p is considered a special order-preserving function. Under this aspect, we derive several results that are relevant for the organization of controlled teaching/learning processes. First, we briefly take up and expand on the idea mentioned previously that a competence space may be seen as a curriculum structure for planning competence-based instruction. Then, several questions are discussed concerning the relations of "learning paths" on the competence and on the performance level.

[8] See Fig. 1 for example: Responsible for the transition from performance state *abc* to state *abce* is the elementary competency $5 \in \mathcal{E}$, *independent* of *which* competence state in $h^{-}(abc) = \{13, 23, 123\}$ underlies performance state *abc*; the elementary competency 5 is what allows the additional solving of problem *e*.

Paths of competence states

It is easily recognized that a competence space $(\mathcal{K}, \cup)$, when considered as a partially ordered set $(\mathcal{K}, \subseteq)$ of competence states, can be regarded as the union of all $\subseteq$-chains in $(\mathcal{K}, \subseteq)$. These chains of competence states suggest an interpretation as *learning paths* or, more precisely, as *paths of learning states*.[9] To maintain the distinction between *competence* and *performance*, we speak of *paths of competence states*; accordingly, learning paths on the level of performance are called *paths of performance states*.

However, such a representation of a partially ordered set $(\mathcal{K}, \subseteq)$ through the set of *all* paths of competence states is, in general, rather large, and formally not very useful. At this point, the didactical perspective of instruction based on a curriculum motivates the restriction to those paths of competence states that provide a most fine-grained subdivision of the curriculum in the smallest possible steps. Such a path is a *maximal chain* in the sense that no state can be added to or inserted in the chain without violating the chain-property (linearity). Apparently, the partially ordered set $(\mathcal{K}, \subseteq)$ is still uniquely determined by this family of all *maximal paths of competence states* (in the sense of all maximal chains on $\mathcal{K}$).[10]

To conclude, each partially ordered set $(\mathcal{K}, \subseteq)$ of competence states may be looked at as the union of a family of (maximal) chains that are reasonably interpretable as (maximal) paths of learning states offering different ways of passing through a curriculum; in a structurally analogous way a partially ordered set $(\mathcal{P}, \subseteq)$ of performance states may be understood as the union of a family of (maximal) paths of performance states that may be regarded as empirically observable sequences of learning steps.

Preserving paths of competence states

Recalling our main topic of consideration, namely showing how union-stable diagnostics may contribute to goal-oriented adaptive instruction, the interpretation developed above now gives rise to some inevitable questions concerning the relations of paths on the competence and on the performance level. Given a union-stable diagnostic, the following questions are of central interest:

(1) Is a (maximal) path of competence states *always* represented by a (maximal) path of performance states?

(2) Is a (maximal) path of performance states *always* a representation of a (maximal) path of competence states?

[9] The idea of relating a knowledge structure to a collection of weak orders that are interpreted as "learning paths" is introduced and elaborated in Koppen (1989).

[10] We do not pursue the question here of how a competence space may be equivalently represented by a family of maximal paths. Clearly, a given family of maximal paths, in order to determine a competence space uniquely, would have to satisfy additional conditions that ensure the union-stability of the resulting competence structure.

The answer to the first aspect of question (1), concerning the representation of paths of competence states in general, is positive.

PROPOSITION 7.7 In a union-stable diagnostic $(\mathcal{E}, \mathcal{K}, \mathcal{A}, \mathcal{P}, k, p)$, any path of competence states is mapped by the representation function p to a path of performance states.

PROOF: Because p is order-preserving (see Lemma 7.1), each path of competence states, $\mu_1 \subsetneq \mu_2 \subsetneq \ldots \subsetneq \mu_{n-1} \subsetneq \mu_n$ (of length n, $n \in \mathbb{N}$) in $(\mathcal{K}, \subseteq)$ is mapped to $p(\mu_1) \subseteq p(\mu_2) \subseteq \ldots \subseteq p(\mu_{n-1}) \subseteq p(\mu_n)$, which is a path of performance states in $\mathcal{P}$.[11] □

We will now examine whether the preserving-property can be transferred to *maximal* paths. However, the following example demonstrates that even in a union-stable diagnostic not necessarily each *maximal* path of competence states is represented by a *maximal* path of performance states.

EXAMPLE: Consider Example 7.4 (see also Fig. 1). The maximal path of competence states $\emptyset \subsetneq 3 \subsetneq 23 \subsetneq 123 \subsetneq 1234 \subsetneq 12345$ is mapped to $\emptyset \subsetneq bc \subsetneq abc \subsetneq abcde$, which is *not* a maximal path of performance states, because the state $abce$ could be inserted. ♣

In contrast to this negative result, question (2) can be answered *positively* for the case of union-stable diagnostics.

PROPOSITION 7.8 In a union-stable diagnostic $(\mathcal{E}, \mathcal{K}, \mathcal{A}, \mathcal{P}, k, p)$, any (maximal) path of performance states is a representation of a (maximal) path of competence states.

PROOF: In the following, let $(\mathcal{K}/\mathit{Ker}(p), \sqsubseteq)$ be the partially ordered congruence partition on $\mathcal{K}$ modulo $\mathit{Ker}(p)$.

(1) Let $Z_1 \subsetneq Z_2 \subsetneq \ldots \subsetneq Z_{n-1} \subsetneq Z_n$, for some $n \in \mathbb{N}$, be a path of performance states. Because the representation function p is surjective, there exist competence states $\mu_1, \mu_2, \ldots, \mu_{n-1}, \mu_n \in \mathcal{K}$, with $Z_i = p(\mu_i)$ for $i = 1, \ldots, n$. Because of Proposition 7.6 (2), $[\mu_1] \sqsubsetneq [\mu_2] \sqsubsetneq \ldots \sqsubsetneq [\mu_{n-1}] \sqsubsetneq [\mu_n]$ is a $\sqsubseteq$–chain on $\mathcal{K}/\mathit{Ker}(p)$. Now, because of Proposition 7.5 (4), for $\kappa, \lambda \in \mathcal{K}$ holds $[\kappa] \sqsubseteq [\lambda] \iff \max[\kappa] \subseteq \max[\lambda]$, thus,

$$\max[\mu_1] \subsetneq \max[\mu_2] \subsetneq \ldots \subsetneq \max[\mu_{n-1}] \subsetneq \max[\mu_n]$$

is a path of competence states that is mapped to the given path of performance states.

[11] Note that the proof only uses the conditions that $\mathcal{K}$ and $\mathcal{P}$ be partially ordered sets, related by an order-preserving function p.

(2) Assume now that the path of performance states given in (1) is *maximal*, beginning with the state $\emptyset$ and running to the top $\mathcal{A} \in \mathcal{P}$. Then, using Proposition 7.6 (2), also the assigned $\sqsubseteq$−chain on $\mathcal{K}/\ker p$ from (1) is maximal, beginning with the bottom element $[\mu_1] \equiv [\emptyset]$, and running to the top element $[\mathcal{E}]$. We now inductively construct a maximal path of competence states running through this $\sqsubseteq$−chain:

(i) Consider the least element $[\mu_1]$ of the chain. Clearly, there exists a $\subseteq$−path from $\emptyset$ to max $[\mu_1]$ in $[\mu_1]$ (for example the path $\emptyset \subseteq \max\,[\mu_1]$). Therefore, there exists also a *maximal* $\subseteq$−path $\emptyset \subseteq \mu_{1_1} \subseteq \cdots \subseteq \mu_{1,r_1} \equiv \max\,[\mu_1]$, with all states being in $[\mu_1]$, because each congruence class in $\mathcal{K}/\mathit{Ker}\,(p)$ is convex relative to $\subseteq$ in accordance with Proposition 7.4 (2).

(ii) Now, let for the chain $[\mu_1] \sqsubseteq \ldots \sqsubseteq [\mu_{i_0-1}]$ $(i_0 \in \{2,\ldots,n\})$ of congruence classes $\emptyset \subseteq \ldots \subseteq \max\,[\mu_{i_0-1}]$ be a maximal path within the classes $[\mu_1] \ldots, [\mu_{i_0-1}]$. According to Proposition 7.5 (3), for max $[\mu_{i_0-1}]$, there exists at least one state $\nu \in [\mu_{i_0}]$, with $\max\,[\mu_{i_0-1}] \subseteq \nu$; let $\mu_{i_0,1}$ be a *minimal* state from $[\mu_{i_0}]$ with this property. Then, a maximal $\subseteq$−chain $\mu_{i_0,1} \subseteq \mu_{i_0,2} \subseteq \cdots \subseteq \mu_{i_0,r_{i_0}} \equiv \max\,[\mu_{i_0}]$ can be found, the members of which are all contained in $[\mu_{i_0}]$ (again, using the convexity of each class). Therefore,

$$\emptyset \subseteq \ldots \subseteq \max\,[\mu_1] \subseteq \ldots \subseteq \max\,[\mu_{i_0-1}] \subseteq$$
$$\subseteq \mu_{i_0,1} \subseteq \cdots \subseteq \mu_{i_0,r_{i_0}} \equiv \max\,[\mu_{i_0}]$$

is a maximal path of competence states.

Pursuing this method of construction, one finally obtains a *maximal* path of competence states, starting from $\emptyset$ and arriving at $\max\,[\mu_n] \equiv \mathcal{E}$, that is mapped to the given maximal path of performance states. □

It should be emphasized that the proof for Proposition 7.8 is a *constructive* one. It explicitly yields for any given maximal path of performance states a maximal path of competence states from which the former is a representation on the performance level. Nevertheless, the method utilized for construction is not the only one possible as indicated by the following example.

EXAMPLE 7.5 In the union-stable diagnostic from Example 7.4 (see also Fig. 1), the maximal path of performance states

$$\emptyset \subseteq bc \subseteq abc \subseteq abce \subseteq abcde$$

is a representation of the following two maximal paths of competence states

$$\emptyset \subseteq 3 \subseteq 13 \subseteq 123 \subseteq 1235 \subseteq 12345$$
$$\emptyset \subseteq 3 \subseteq 23 \subseteq 123 \subseteq 1235 \subseteq 12345\,;$$

these two maximal paths are obtained according to the construction method from the proof of Proposition 7.8. However, in addition to these paths there are other paths of competence states mapped onto the considered maximal path of performance states, e.g.

$$\emptyset \subseteq 3 \subseteq 13 \subseteq 135 \subseteq 1235 \subseteq 12345$$
$$\emptyset \subseteq 3 \subseteq 23 \subseteq 235 \subseteq 1235 \subseteq 12345$$
$$\emptyset \subseteq 3 \subseteq 13 \subseteq 135 \subseteq 1345 \subseteq 12345$$
$$\emptyset \subseteq 3 \subseteq 23 \subseteq 235 \subseteq 2345 \subseteq 12345.$$

♣

At this point, our questions have been completely answered. To summarize, within the framework of a union-stable diagnostic principally *two possible ways of planning instruction sequences* may be followed:

1. A *competence-based strategy* may be applied that, dependent on the (possibly not uniquely) diagnosed competence state, selects a suitable next learning goal in the competence space, teaches the uniquely determined elementary competencies, and tests the arrival at the new learning goal by observing the solution behavior on corresponding problems. In this manner a path of competence states as a goal-orientiation for instruction would be adaptively followed, evaluated by observable transitions along a corresponding path of performance states.
2. A *performance-based strategy* would follow a selected path of performance states directed by the plan that, gradually, additional problems along that path should become solvable; such a performance orientation would be possible because for any path of performance states there exists a path of corresponding competence states that orders the required elementary competencies. This strategy can even be didactically optimized by utilizing *maximal* paths on the performance as well as on the competence level.

To conclude, the considerations and results of this chapter should demonstrate that the competence-performance approach to the knowledge-structure theory and, especially, the concept of a union-stable diagnostic may provide an interesting contribution to research on and practical application of planning and controlling of adaptive, goal-directed learning and instruction.

ACKNOWLEDGEMENTS

The research reported in this paper was supported by Grant Lu 385/1 of the Deutsche Forschungsgemeinschaft to J. Lukas and D. Albert at the University of Heidelberg.

REFERENCES

Doignon, J.-P., & Falmagne, J.-C. (1998). *Knowledge spaces.* Berlin: Springer.

Falmagne, J.-C., Koppen, M., Villano, M., Doignon, J.-P., & Johannesen, L. (1990). Introduction to knowledge spaces: How to build, test, and search them. *Psychological Review, 97,* 201–224.

Koppen, M. (1989). *Ordinal data analysis: Biorder representation and knowledge spaces.* Katholieke Universiteit te Nijmegen.

Korossy, K. (1993). *Modellierung von Wissen als Kompetenz und Performanz. Eine Erweiterung der Wissensstruktur-Theorie von Doignon und Falmagne.* [Modeling knowledge as competence and performance. An extension of Doignon and Falmagne's theory of knowledge spaces.] Unpublished doctoral dissertation, University of Heidelberg, Germany.

8

Structure and Design of an Intelligent Tutorial System Based on Skill Assignments

Dietrich Albert
Karl-Franzens-Universität Graz

Martin Schrepp
Ruprecht-Karls-Universität Heidelberg

Since the beginning of research in artificial intelligence several attempts have been made to construct intelligent tutorial systems (ITS). Such an ITS consists in general of a representation of the knowledge in a special domain, a diagnostic procedure to determine the knowledge of a student working with the system, teaching material, and a procedure for adaptive teaching. This chapter demonstrates how an extension of the theory of knowledge spaces can be used for the design of a domain-independent ITS. The main components of this ITS are a representation of the skills necessary in the knowledge domain and their dependencies as a surmise system, a set of questions related through a skill assignment to the skill states used for knowledge diagnosis, and a rule that relates skill states to teaching operations. The ITS is adaptive with respect to the consideration of the prior knowledge a student possess and with respect to the learning speed of a student. The strict formalized description of the systems components and their interactions during the teaching process guarantees that an ITS with the described properties can be implemented easily.

INTRODUCTION

One of the research topics discussed in artificial intelligence is the approach to use computers in education for diagnosing knowledge (for example Brown & Burton, 1978) or teaching (e.g., Mandl & Hron, 1985; Okamoto & Matsuda, 1993; Sleeman & Brown, 1982).

Intelligent tutorial systems (ITS) differ from other forms of computer-assisted-instruction (e.g., drill-programs, electronic textbooks, or simulation programs) in being "intelligent" with respect to two criteria. First, the system must be able to diagnose the status of the student's knowledge in the underlying knowledge domain. Second, the tutorial strategy must be adaptive to the prior knowledge of the student and must be able to improve the learner's knowledge.

There exists a variety of architectures for ITS in different knowledge domains. Most of the constructed systems work in highly formalized domains, like subdisciplines of mathematics, for example, symbolic logic (Matsuda & Okamoto, 1992), arithmetic (Takeuchi & Otsuki, 1994), fraction calculation (Kondo, Watanabe, Takeuchi, & Otsuki, 1990), integration (Kimball, 1982), physics (Ploetzner & Spada, 1993), language acquisition (Kunichika, Takeuchi, & Otsuki, 1994), or the usage of software systems or programming languages (Heines & O'Shea, 1985; Yasuda & Okamoto, 1991).

Most of these architectures are highly specialized in their specific knowledge domain and it is, therefore, not possible to use them in other domains.

There is a wide agreement in literature that an ITS must contain at least four basic components, that interact during the teaching process. These components are a *knowledge base*, a *student model*, a *teaching component*, and a *diagnostic component*. Existing ITS differ in the way in which these components are conceptualized, especially in their in general domain specific student models.

The knowledge base contains all relevant knowledge of the domain for which the ITS is constructed. This component represents the knowledge of an expert in the domain and is often realized as an expert system.

The student model is a representation of the cognitive abilities or skills of students in the knowledge domain. It consists, in general, of a set of states. Each state represents a special state of knowledge in which a student may possibly be during the learning process, that is during his or her interaction with the ITS. The learning process is represented by a sequence of states of the student model.

The teaching component consists of teaching materials (instructions, exercises, or demonstrations) and a rule that determines for each state of the student model, which part of this material is relevant for a student who is in that particular state.

The state of knowledge of a student is, in general, not directly observable because the student model contains hypothetical assumptions about necessary cognitive abilities or cognitive procedures. The diagnostic component is used to infer this state from the interaction of the student with the ITS.

The goal of the teaching process is to enable the student to learn all relevant information of the knowledge domain (or, at least, a sufficient large subset of this information), so that her or his knowledge after the teaching process terminates is more or less equal to the knowledge contained in the knowledge base of the ITS.

We want to show how the theory of knowledge spaces (Doignon & Falmagne, 1985, 1998) and skill assignments (Doignon, 1994a; Lukas & Albert, 1993; Korossy, 1993) can be used to develop a general architecture of an ITS. This general architecture is domain independent. Therefore it can be used for the concrete construction of ITS's in different knowledge domains.

Two properties of the theory of knowledge spaces and skill assignments are especially important for this approach. First, the theory of knowledge spaces allows the formulation of effective and adaptive procedures for the diagnosis of knowledge (see Falmagne & Doignon, 1988; Doignon, 1994b). Second, skill assignments allow a precise description of the connection between observable behavior in problem solving and underlying cognitive abilities or skills (Doignon, 1994a; Korossy, 1993, 1997).

In the following section, we describe the basic ideas of an ITS based on the theory of knowledge spaces and skill assignments. Then, we show how this basic system can be generalized by incorporating some important new components. In the last section possibilities for further developments are sketched and open questions leading to further research are discussed.

BASIC IDEAS OF AN ITS BASED ON SKILL ASSIGNMENTS

In this section, we develop the basic architecture of an ITS based on skill assignments. This basic architecture is very simple and therefore limited in its applicability because the main goal of this section is to clarify the basic ideas underlying our approach.

For the formal description of our basic architecture we have to introduce some notations. We write in the following Pow(X) for the power set, the set of all subsets of a set X. Seq(X) denotes the set of all tuples of elements of a set X with Seq(X) $:= \{(x_1, \ldots, x_n) \mid n \in N \wedge x_1, \ldots, x_n \in X\}$.

We divide the components of our ITS into *statical* and *dynamical* components. A component is called dynamical if it is updated during the teaching process, thus being adaptive to the student's behavior, and statical if this is not the case. We begin with the description of the statical components.

Knowledge Base

The knowledge necessary in the domain is represented through a set S of *skills*. Dependencies between these skills are described by a surmise function $\sigma : S \rightarrow \text{Pow}(\text{Pow}(S))$. A surmise function assigns to each skill the different sets of prerequisites of this skill. We interpret $\sigma(s) = \{S_1, \ldots, S_n\}$ as "Every student has to master all skills from at least one of the subsets $S_1, \ldots, S_n$ of S to be able to reach mastery of s". The surmise function σ is used during the teaching process to determine an optimal learning path. The surmise system (S, σ) can be considered as the *knowledge base* of the system.

Student Model

Given S and σ, we can define the *student model* $\mathcal{S}$ by:

$$\mathcal{S} := \{S' \subseteq S \mid \forall s \in S' \; \exists S'' \in \sigma(s) \; (S'' \subseteq S')\},$$

$\mathcal{S}$ is the set of all subsets of S that contain with every skill s at least one set S'' of prerequisites of s. $\mathcal{S}$ is the set of all subsets of S consistent with the dependencies of skills described by σ. Therefore, the knowledge of every student working with the system can be described by an element of $\mathcal{S}$, if we presuppose that S contains all the skills relevant[1] in the knowledge domain and that σ describes the dependencies between these skills correctly. If the knowledge of a student is described by $S' \in \mathcal{S}$ we call S' the *competence state* of that student.

Diagnostic Component

The skills in S are hypothetical and are not directly observable constructs. Therefore, our system must contain a possibility to infer the skills a student possess, the student's competency, from his or her solving-behavior.

Let Q be a set of questions or problems in the underlying knowledge domain. We assume in the following that the questions in Q are especially constructed for the diagnosis of the skills in S. Thus, we can assume that the mastery of all skills in S is sufficient for the ability to solve all questions in Q. None of the questions in Q require a skill $s \notin S$. The systematical construction of questions for a given set of skills is discussed for example in Lukas and Albert (1993), Held (1993), Albert, Schrepp, and Held (1994), or Korossy (1993, 1997).

Because we want to diagnose the skills a student possesses from her or his solving behavior, we have to relate the skills in S to the ability of students to solve the questions in Q. This is done by a *skill assignment* $\alpha : Q \rightarrow \text{Pow}(S)$. For a

[1] Relevant skills are the skills that form the specific body of knowledge in the underlying knowledge domain. The ability to add two integers is an example of such a relevant skill in the domain of elementary arithmetic. General skills that are necessary for understanding the instructions or problems of the domain, such as the ability to read, are presupposed and not considered as members of S.

question q, we interpret $\alpha(q) = \{S_1, \ldots, S_n\}$ as "A student answering correctly to question q must master all skills from at least one of the sets $S_1, \ldots, S_n$". Because of this interpretation we can assume that the states $S_1, \ldots, S_n$ of the student model contained in $\alpha(q)$ are minimal with respect to $\subseteq$, for all $S_i, S_j \in \alpha(q)$ we have $S_i \not\subseteq S_j$. For a more detailed discussion of the dependency between the skills possessed by a student and her or his solving behavior, see Korossy (1993, 1997).

The skill assignment determines for each state $S' \in \mathcal{S}$ of the student model the answer behavior $\beta(S')$ consistent with this state by

$$\beta(S') := \{q \mid q \in Q \wedge S' \in \alpha(q)\},$$

$\beta(S')$ is the set of all questions in Q that can be answered correctly by a student with competency S'.

Teaching Component

To allow teaching operations, the system must contain materials that can be presented to students in order to improve their knowledge. Let I be a set of instructions, like examples, facts, informational texts, or exercises. We represent I as a union of pairwise disjoint sets $I = I_1 \cup \ldots \cup I_n$, with $I_l \cap I_m \neq \emptyset$ if and only if $l = m$. Each of these sets I_l represents a special type of instruction.

The teaching operations should be adaptive to the prior knowledge of a student. Therefore, we have to relate them to the skills in S. This relation is established by a function $\delta : S \times \text{Pow}(S) \rightarrow \text{Seq}(I)$, which relates skills and their prerequisites to an instructional sequence. We interpret $\delta(s, S') = (i_1, \ldots, i_n)$ for $n \in \mathbf{N}$ as "The instructional sequence $i_1, \ldots, i_n$ should be presented to a student who masters all skills in $S' \setminus \{s\}$ and who does not master s". (I, δ) can be considered as the *teaching component* of the system.

Hypothetical Student States

To enable an adaptive teaching strategy, the system must also contain dynamical components, that is components that change during the interactions of a student with the system.

The actual information of the system about the knowledge of a student working with it can be represented as a subset $\mathcal{M}$ of $\mathcal{S}$. This subset $\mathcal{M}$ consists of all states of $\mathcal{S}$ that are consistent with the previous answers of the student on the presented questions. We call $\mathcal{M}$ the *set of hypothetical student states*. If $q_1, \ldots, q_n \in Q$ are the questions already answered correctly by a student and $q'_1, \ldots, q'_m$ the questions answered incorrectly by that student we have:

$$\mathcal{M} = \{S' \in \mathcal{S} \mid \forall i = 1, \ldots, n\ (q_i \in \beta(S')) \wedge \forall i = 1, \ldots, m\ (q'_i \notin \beta(S'))\}.$$

Storage of Information

The interactions between student and system must be stored to use them for teaching decisions. For example, which questions the student has already answered must be stored and also which information was already presented to him or her. This is done in the history H of the teaching process. H is a list of all previously presented questions, obtained answers, and presented instructions. Thus, H is an element of $\text{Seq}((Q \times R) \cup I)$, where $R = \{\text{correct}, \text{false}\}$. For example $H = ((q_1, r_1), (q_2, r_2), i_1, i_2, i_3, (q_3, r_3), \ldots)$ means that first question q_1 was presented and answer r_1 was obtained, second question q_2 was presented and answer r_2 was obtained. Afterwards, instructions i_1, i_2, i_3 were presented to the student, followed by a presentation of question q_3, on which answer r_3 was obtained, and so on.

Teaching Process

Up to this point, we have described only the components of the systems architecture. Now, we describe how these components interact during the teaching process. The teaching process can be considered as an interaction between five modes of the system.

The first mode is the START mode. Here the system sets $\mathcal{M} = \mathcal{S}$ because, up to this point, no information about the students knowledge is obtained. Then, the system changes into the second mode, called DIAGNOSIS mode.

The DIAGNOSIS mode consists of three steps. In the first step, the system tries to determine a question q, for which an obtained answer guarantees an optimal reduction of the set of hypothetical student states consistent with the obtained answer of the student. This can be done by the half-split procedure in the diagnostic algorithm described in Falmagne and Doignon (1988).

For the set $\mathcal{M}$ of hypothetical student states and a question $q \in Q$, we define $\mathcal{M}_{q,1} \subseteq \mathcal{M}$ as the set of all elements of $\mathcal{M}$ that are consistent with the fact that the student solves question q and $\mathcal{M}_{q,0} \subseteq \mathcal{M}$ as the set of all elements of $\mathcal{M}$ that are consistent with the fact that the student fails in solving q. Formally, we have $\mathcal{M}_{q,1} = \{M \in \mathcal{M} \mid q \in \beta(M)\}$ and $\mathcal{M}_{q,0} = \{M \in \mathcal{M} \mid q \notin \beta(M)\}$. The sets $\mathcal{M}_{q,1}$ and $\mathcal{M}_{q,0}$ are pairwise disjoint, that is we have $\mathcal{M}_{q,1} \cap \mathcal{M}_{q,0} = \emptyset$ and we have $\mathcal{M}_{q,0} \cup \mathcal{M}_{q,1} = \mathcal{M}$.

We define a question $q' \in Q$ as being *optimal for diagnosis* if the value of $\mid \text{card}(\mathcal{M}_{q',1}) - \text{card}(\mathcal{M}_{q',0}) \mid$ is minimal over all $q \in Q$. For an optimal question q' the cardinality of the sets $\mathcal{M}_{q',1}$ and $\mathcal{M}_{q',0}$ is optimal balanced, that is the sets $\mathcal{M}_{q',1}$ and $\mathcal{M}_{q',0}$ differ in cardinality as little as possible.

The described procedure can be regarded as the *diagnostic component* of the system. Notice that a question already answered (correct or incorrect) by the student can not be optimal as long as the set $\mathcal{M}$ of hypothetical student states contains at least two elements.

In the second step, an optimal question q, determined by the first step, is presented to the student. If the answer of the student is $r \in \{0, 1\}$ (where 1 represents "correct" and 0 "false") then $\mathcal{M}$ is replaced by $\mathcal{M}_{q,r}$.

The third step tests if there exists a skill s not mastered by the student, (formally: $\forall M \in \mathcal{M}(s \notin M)$), for which a set of prerequisites is completly mastered, (formally: there exists a $S' \in \sigma(s)$ with $\forall M \in \mathcal{M}(S' \setminus \{s\} \subseteq M)$). If this is the case, then the system changes to the TEACH mode. If this is not the case, the system continues with the first step of the DIAGNOSIS mode.

The TEACH mode starts if the DIAGNOSIS mode has detected a skill s not mastered by a student who masters for a set S' of prerequisites of s all skills from $S' \setminus \{s\}$. Formally, this means that $\forall M \in \mathcal{M}(s \notin M)$ and $\forall M \in \mathcal{M}(S' \setminus \{s\} \subseteq M)$.

It is possible that the DIAGNOSIS mode has detected more than one skill with this properties. If this is the case, the system has to decide which of these skills should be taught first. We discuss this problem in the following section. For the moment, it is sufficient to assume that the decission is made by chance.

The TEACH mode then presents the instructional sequence $\delta(s, S')$ to the student. After the presentation of this instructional sequence the system changes to the TEST mode.

The TEST mode consists of three steps. In the first step the system checks if the student has reached mastery of the last skill s taught by the TEACH mode. Therefore, a question is presented for which skill s is necessary and only skills from S' are required, where S' denotes the set of prerequisites to s detected in the third step of the DIAGNOSIS mode. Formally, this means that the system presents a question q with $\forall S'' \in \alpha(q)(s \in S'' \wedge S'' \subseteq S')$.

In the second step, the system checks if the presented question is answered correctly by the student, indicating that he or she had learned the required skill. If the student fails in answering correct, that is if $r \neq 1$, the system goes back to the TEACH mode and starts the presentation of $\delta(s, S')$ again. If the student gives the correct answer to the presented question $\mathcal{M}$ is replaced by $\{M \cup \{s\} \mid M \in \mathcal{M} \wedge M \cup \{s\} \in \mathcal{S}\}$ and the third step of the TEST mode checks if $\mathcal{M} = \{S\}$, indicating that the student masters all skills required in the domain. If this is the case the system changes to the END mode, if not the system changes to the third step of the DIAGNOSIS mode.

The END mode informs the student that he or she had reached the learning goal and finishes the process.

The teaching process is depicted in Fig. 1 as a flow chart.

GENERALIZATIONS OF THE BASIC SYSTEM

In the previous section, we introduced an architecture for an ITS based on the theory of knowledge spaces. Because our main goal in this section is to work out the general ideas underlying an application of knowledge space theory in the field

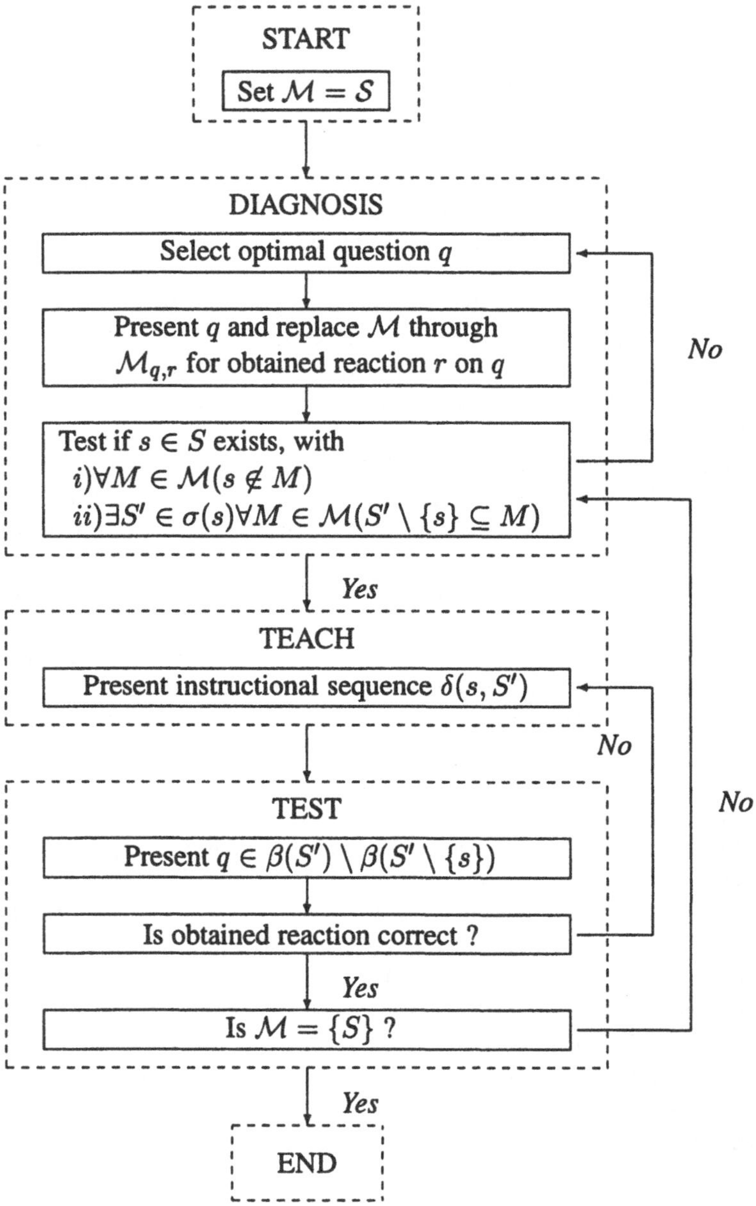

FIG. 1. Teaching process described as an interaction between the five modes of the system.

of computerized teaching, this basic architecture is very simple and is therefore restricted in its applicability. In this section, we try to clarify the limitations of the basic architecture and work out methods to overcome them. So, at the end of this section, we are able to formulate a generalization of our basic architecture.

In our basic architecture, we have considered only two possible answer types of a student to a presented question by assuming that such an answer is either correct or incorrect. This approach has the advantage of simplicity, but does not use the whole information about the student's competency contained in the obtained answer.

To use this information completely we have to introduce the answer in the skill assignment and clarify the role of the answer in the update of the set of hypothetical student states $\mathcal{M}$ within the first and second step of the DIAGNOSIS mode.

Therefore, we introduce the set R of all possible answers (possible inputs to the system) a student can give to the questions in Q. We relate the skills in S to the questions in Q and possible answers in R by a *skill assignment*[2] $\alpha : Q \times R \to \text{Pow}(S)$. The interpretation of $\alpha(q,r) = \{S_1, \dots, S_n\}$ is "A student who answers r to question q must at least master all skills from one of the sets $S_1, \dots, S_n$". As in the previous section, we can assume that the sets $S_1, \dots, S_n$ in $\alpha(q,r)$ are minimal with respect to $\subseteq$, i.e. $\forall S_i, S_j \in \alpha(q,r)\ (S_i \not\subseteq S_j)$.

As in the previous section, the skill assignment α determines for each state $S' \in \mathcal{S}$ the answer behavior $\beta(S')$ consistent with this state by

$$\beta(S') = \{(q,r) \mid q \in Q \wedge S' \in \alpha(q,r)\},$$

$\beta(S')$ is the set of all tuples of questions and obtained answers that can be produced by a student with competency S'.

Now, we have to generalize the half-split procedure in the first step of the DIAGNOSIS mode. For the set $\mathcal{M}$ of hypothetical student states and a question $q \in Q$, we define $\mathcal{M}_{q,r} \subset \mathcal{M}$ as the set of all elements of $\mathcal{M}$, that are consistent with the fact that the student gives answer r on question q, $\mathcal{M}_{q,r} = \{M \in \mathcal{M} \mid (q,r) \in \beta(M)\}$. We define a question $q' \in Q$ as being *optimal for diagnosis* if the value $\mid \max_{r,r' \in R} (\text{card}(\mathcal{M}_{q',r}) - \text{card}(\mathcal{M}_{q',r'})) \mid$ is minimal over all $q \in Q$. This means that for the optimal question q' the cardinality of the sets $\mathcal{M}_{q',r}$ is for all possible answers r optimal balanced, that is the sets $\mathcal{M}_{q,r}$ differ in cardinality as little as possible. The sets $\mathcal{M}_{q',r}$ must not be pairwise disjoint, we can have $\mathcal{M}_{q',r} \cap \mathcal{M}_{q',r'} \neq \emptyset$ for $r \neq r'$, and they can be empty, that is we can have $\mathcal{M}_{q',r} = \emptyset$ for some r.

As before in the second step of the DIAGNOSIS mode, an optimal question is presented to the student and if the answer is $r \in R$, then $\mathcal{M}$ is replaced by $\mathcal{M}_{q',r}$.

[2] In the following we denote concepts that are proper generalizations of concepts introduced earlier by the same symbols as in the previous section. Since, for example, the generalized concept of a skill assignment is for $R = \{\text{correct}, \text{incorrect}\}$ identical with the concept of a skill assignment as introduced in the previous section, it is denoted by the same greek letter α.

To make things easy, in the formulation of the basic architecture we have drawn implicitly the unrealistic assumption that a student answers always accordingly to her or his competency described by a state of the student model. This assumption ensures that our rules governing the update of $\mathcal{M}$ in the DIAGNOSIS mode do not end with an empty $\mathcal{M}$. Remember that the teaching process starts with $\mathcal{M} = \mathcal{S}$ and that after each obtained answer r of a student to a question q, the set $\mathcal{M}$ is replaced by $\mathcal{M}_{q,r} \subseteq \mathcal{M}$.

But if, as a result of a lucky guess or a careless error, the students answer r to question q is not consistent with one of the states in $\mathcal{M}$, that is with the students previous interactions with the system, we can receive an empty $\mathcal{M}_{q,r}$. In this case, we have to find a new rule for the update of $\mathcal{M}$ in step two of the DIAGNOSIS mode.

We distinguish between two cases. If $\mathcal{M}_{q,r} \neq \emptyset$ we replace just as before $\mathcal{M}$ by $\mathcal{M}_{q,r}$. If conversely $\mathcal{M}_{q,r} = \emptyset$ we replace $\mathcal{M}$ by $\mathcal{M} \cup \{S' \in \mathcal{S} \mid \exists M \in \mathcal{M} \ (\mid (M \setminus S') \cup (S' \setminus M) \mid = 1)\}$, enlarging $\mathcal{M}$ by considering all states of the student model that differ only in one skill from the states in $\mathcal{M}$. The enlargement of $\mathcal{M}$ is necessary because we have to consider the possibility that the obtained answer r on q is in accordance with the competency of the student and $\mathcal{M}_{q,r}$ is empty as a result of a lucky guess or careless error to one of the previous questions, that leaded to an elimination of the correct state of the student.

To ensure that a wrong assumption concerning the students' competency, for example, resulting from a careless error, can be corrected, we have to assume that $\mathcal{S}$ is *well–graded* (Falmagne & Doignon, 1988). This means that for two arbitrary states $S', S'' \in \mathcal{S}$, there exists a chain $S' = S_1 \subset S_2 \subset \ldots \subset S_n = S''$ of states in $\mathcal{S}$ with card$(S_i \setminus S_{i-1}) = 1$ for $i = 2, \ldots, n$. Each S_i contains exactly one skill more than his predecessor in the chain.

This assumption is necessary because we enlarge $\mathcal{M}$ in the case of an error in diagnosis only by the states that differ in only one element from the states in $\mathcal{M}$. If in a not well–graded $\mathcal{S}$ one state S' differs in at least two skills from all other states, this state can not be reached by the described rule if it is once eliminated from $\mathcal{M}$, for example as a result of a careless error.

One important restriction of the basic architecture is that, given a skill s not mastered by a student who masters all skills from a set $S' \setminus \{s\}$, where S' is a set of prerequisites of s, there is only one sequence $\delta(s, S') = (i_1, \ldots, i_n)$ of instructions that lead to the mastery of s by the student. Generally one would expect many such instructional sequences having all the same intended effect on the students' knowledge. For example, a missing skill may be taught alternatively by presenting some examples, some instructional texts, or a mixture of both. Which of these alternatives is best for a particular student may depend on the student's personality.

We include the possibility of alternative instructional sequences by assigning to a skill s and a set S' of prerequisites of s a set of instructional sequences through a function δ. So we regard δ as a function $\delta : S \times \text{Pow}(S) \rightarrow \text{Pow}(\text{Seq}(I))$. The interpretation of $\delta(s, S') = \{I_1, \ldots, I_n\}$ is "To a student mastering all skills from S' and not mastering s, one of the instructional sequences $I_1, \ldots, I_n$ should be presented to reach mastery of s".

Because we had assumed in our basic architecture that $\delta(s, S')$ contains only one instructional sequence, we must assume that this instructional sequence is repeated until the student masters skill s. If the TEST mode finds that a student has not learned the last skill taught by the TEACH mode by presenting the sequence $\delta(s, S')$, this sequence is presented again. This may be problematic for two reasons. First, the instructional sequence $\delta(s, S')$ may be inappropriate for the student, so that a repetition of this sequence makes no sense. Second, a repetition of informations already presented may lead to a motivational decrease.

We can overcome this limitation by assuming that solely such instructional sequences can be presented to a student that have not been presented before as long as $\delta(s, S')$ contains at least one informational sequence not already presented. Formally, this means that to a student, who had already received the instructional sequences $I_1, \ldots, I_n$ (these sequences are stored in the history of the teaching process), only instructional sequences from $\delta(s, S') \setminus \{I_1, \ldots, I_n\}$ can be presented if this set is not empty. If $\delta(s, S') \setminus \{I_1, \ldots, I_n\}$ is empty, then an arbitrary element of $\delta(s, S')$ determined by chance should be presented again to the student.

Now the problem arises which of these sequences $I_1, \ldots, I_n \in \delta(s, S')$ should be presented to a student during the TEACH mode of the system. This can be solved by a *choice rule* γ which chooses one of these sequences with respect to the previous interactions between system and student.

Such a choice rule can be implemented in several ways. In the following, we discuss one such possibility. We divide the set $\text{Seq}(I)$ of instructional sequences into a finite number of pairwise disjoint subsets $\mathcal{C}_1, \ldots, \mathcal{C}_n$, that is we have $\text{Seq}(I) = \mathcal{C}_1 \cup \ldots \cup \mathcal{C}_n$ and $\mathcal{C}_i \cap \mathcal{C}_j = \emptyset$ for $i \neq j \in \{1, \ldots, n\}$.

These subsets $\mathcal{C}_i$ represent instructional types, for example instruction by examples, or instruction based on texts. For the classification of sequences into types, the description of I as a union of pairwise disjoint sets of different instructions introduced in the previous section can be used.

We can expect that students differ in the type of instruction they prefer and with which they learn best. So the choice rule γ should ensure that a student receives the type of instruction which is most effective for her or him, that is γ should be adaptive to the success of a special instructional type in the previous interactions between system and student.

We define the *effectivity* $\epsilon(\mathcal{C}_i)$ of an instructional type $\mathcal{C}_i$ as the relative frequency of already presented instructional sequences from $\mathcal{C}_i$ leading to a mastery of the intended skill. The effectivity $\epsilon(\mathcal{C}_i)$ can be measured by the TEST mode. So the effectivity ϵ is a function $\epsilon : \{\mathcal{C}_1, \ldots, \mathcal{C}_n\} \times H \rightarrow [0, 1]$.

We use the effectivity for the construction of an adaptive choice rule by assuming that the probability $p(\mathcal{C}_i)$ for the presentation of a sequence $I_i \in \delta(s, S')$

increases with the effectivity $\epsilon(C_j)$ of the instructional type to which I_i belongs. If $I_1 \in C_1, I_2 \in C_2 \in \delta(s, S')$ we should have $p(I_1) \geq p(I_2)$ if and only if $\epsilon(C_1) \geq \epsilon(C_2)$. The decision between members of the same instructional type is drawn by chance as before.

Another limitation of the basic architecture is that the goal of the teaching process is mastery of all skills in S by the student, the teaching process terminates only if $\mathcal{M} = \{S\}$. In a practical application it may be for two reasons necessary to define subgoals of the teaching process.

First, for some students the mastery of a subset of S may be sufficient, for example, if they need only basic knowledge in the domain and do not want to become experts there. Second, the teaching process may have taken too much time for one session, so we have to formulate points where an interuption makes sense.

Such subgoals can be introduced by marking a subset $\mathcal{G}$ of $\mathcal{S}$. An element of $\mathcal{G}$ is called a *subgoal* of the teaching process.

Which elements of $\mathcal{S}$ are suitable as subgoals? A subgoal should be a closed piece of knowledge, all skills in it should be connected. If we interpret the surmise function σ on S as a description of the contentual dependency of the skills the optimal candidates for subgoals are the basis elements of $\mathcal{S}$. The reason is that a basis element B includes with each skill s only such skills, which are either elements of a set of prerequisites of s or for which s itself is an element of one of their sets of prerequisites. Therefore we chose $\mathcal{G}$ as a subset of the basis $\mathcal{B}$ of $\mathcal{S}$.

To make use of the subgoals for the control of the teaching process, we only have to change the third step of the test mode of our system. This step tests if $\mathcal{M} = \{S\}$, if the student masters all skills from S, and changes to the END mode if this is the case. This step can be easily generalized by assuming that it tests if all hypothetical student states include a basis element B, that is if the student masters at least all skills from B. Formally, this means that it is tested if the condition $\exists B \in \mathcal{B} \forall M \in \mathcal{M}\ (B \subseteq M)$ is fulfilled. If this is the case the student is asked if he or she wants to interrupt the process. If the student decides to interrupt, the history H of the teaching process is stored and can be used in the next session to continue the process properly. If the learning goal of a particular student is a proper subset of S, for example, if the student only needs basic knowledge in the domain, the process can be terminated if this special goal is fulfilled. In this case, we have also to ensure that skills not contained in the learning goal $G \subset S$ of the student are not taught, even if the diagnostic procedure finds out that the student does not master these skills. This can be done easily by replacing S by G in all components of the system and in the teaching process. This means that the system "forget" all information about skills in $S \setminus G$ and reacts just as if the set of skills necessary in the domain would be G.

The learning goal of a particular student may be given explicitly, for example, if the system is used by students to fulfill a special course requirement, or asked interactively at the beginning of the teaching process from the student.

The set $\mathcal{G}$ of subgoals of the teaching process can also be used to solve a problem already mentioned in the previous section.

The problem arises if the DIAGNOSIS mode has detected at least two skills s_1 and s_2 not mastered by a student and sets of prerequisites S_1 and S_2 for s_1 respectively s_2 for which all skills from $S_1 \setminus \{s_1\}$ and $S_2 \setminus \{s_2\}$ are mastered by that student. The system has then to decide which of these skills should be taught first. In the previous section, we had assumed that this decision is drawn by chance.

Given a set $\mathcal{G}$ of subgoals we can improve the decision strategy by assuming that skills are taught first that ensure that a subgoal is reached. Only if this rule is not sufficient to decide between two skills, for example if none of these skills leads to a subgoal, the decision is drawn by chance by an implemented random decision strategy. This procedure ensures that the teaching strategy tries to teach first closed pieces of knowledge, i.e., subsets of S in which all skills are contentual connected by σ, instead of unconnected pieces of knowledge.

Another possible extension of our basic architecture is the introduction of the learning speed or learning ability of students into the teaching strategy. Students clearly differ in their ability to learn new material. This difference between students should be considered in an ITS to ensure that the strategy the system uses for the presentation of instructions reflects the learning ability of a student working with the system.

For example, for a student who is able to learn fast it may be optimal to present a number of instructional sequences during one step of the teaching process in order to allow the student to aquire a number of skills during that step. For a student who learns slowly, it seems more appropriate to present only the instructions necessary to learn one missing skill during one step of the teaching process.

The teaching process described in the previous section considers only the second case of the example. This can be seen from the interaction of the TEACH and the TEST mode of the system. Assume a student who does not master skill s but who masters all skills from $S' \setminus \{s\}$, where S' is a set of prerequisites of s. We assumed in the previous section that in such a case, the TEACH mode presents the instructional sequence $\delta(s, S')$ and afterwards the system changes to the TEST mode to ensure that skill s was aquired by the student. Therefore, the procedure described in the previous section seems to be optimal only for students who learn slowly.

Because the learning speed of a student is not known to the ITS before the teaching process starts, it must be measured during this process. This can be done adaptively by the TEST mode. The central idea is that the number of cases in which the student had aquired the skill taught in the previous TEACH step is an indicator for the learning speed of that student.

We measure the learning speed g of a particular student by an integer greater than 0, that is we can have $g = 1, 2, 3, \ldots$. At the beginning of the teaching process we set $g = 1$ for every student. If the TEST mode detects in l successive steps, where l is a fixed level, that all skills[3] taught in the previous step are aquired by the student, then g is replaced by $g + 1$. If the TEST mode detects conversly that a student has not aquired the skills taught in the previous TEACH step and g is greater than 1, then g is replaced by $g - 1$.

Now, we have to describe how the learning speed g influences the choice of skills that will be taught during the TEACH mode. Assume that the DIAGNOSIS mode had detected a skill s not mastered by a student who masters all skills from $S' \setminus \{s\}$, where S' is a set of prerequisites of s. If $g = 1$, then the TEACH mode presents just as before an element $\delta(s, S')$ and changes to the TEST mode. If $g = m > 1$ then the TEACH mode checks if there exists skills $s = s_1, s_2, \ldots, s_m$ with $S' \cup \{s_1, \ldots, s_p\} \in \sigma(s_{p+1})$ for $p = 1, \ldots, m - 1$. This means that $S' \cup \{s_1, \ldots, s_p\}$ is a set of prerequisites of the skill s_{p+1}. If such skills exists, then the system presents instructional sequences from $\delta(s_1, S'), \delta(s_2, S' \cup \{s_1\}), \ldots, \delta(s_m, S' \cup \{s_1, \ldots, s_{m-1}\})$, i.e. gives the student the possibility to aquire all the skills $s_1, \ldots, s_m$ in one step. If such skills does not exist, the system choses the maximal number $p < m$ of skills with the described properties and presents instructional sequences corresponding to these skills. We have to ensure in this case that an aquisition of this $p < m$ skills by the student does not lead to an increase of the value g. So only such cases are considered in the update of g, where the number of skills presented during the TEACH step is equal to the actual value of the learning speed g.

DISCUSSION

The ITS described in this article consists of several structures, for example $S, Q, I, \sigma, \alpha, \delta$, which interact during the teaching process. The design of these structures influences the performance and effectivity of the system. We illustrate this by an example.

The effectivity of the system depends on its ability to diagnose which skills a student does not master. Especially the speed of the diagnosis is very important, since it may be demotivating for a student if a lot of questions will be presented to her or him until a lacking skill is diagnosed and teaching materials are presented. The speed of the diagnosis depends on the functions σ and α. For example a stricter σ results in a faster diagnosis.

Thus, it may be adequate to chose σ as strict as possible, even if this may result in an elimination of some true states from S which occur with low frequency in the intended population. This approach increases the speed of the diagnosis while it decreases its accuracy. Such manipulations of the components must be handled carefully. Their efficiency can be evaluated only in a concrete application.

[3] We assume here that the TEACH mode can teach more than one skill in each step. How this can be realized is described later.

The described ITS contains no procedure for the repetition of already mastered skills. We have not introduced such a procedure since we assumed that the skills are teached in increasing sequence concerning their difficulty[4]. Therefore we can assume that the skills which are teached first, i.e. the easy ones, are practiced again during the aquisition of the skills which require them as prerequisites. But if we assume that the teaching process may be interupted, for example after a subgoal is reached, for a long period of time it seems necessary to introduce a procedure for the repetition of skills. Such a procedure requires a rule for selecting the skills mastered in the previous session which should be repeated at the beginning of the new session. How such a rule can be formulated and integrated into the teaching process is at the moment an open question.

We have formulated our ITS domain independent. For a concrete application of the ITS in a special knowledge domain several steps are necessary. First, the skills relevant in the domain, i.e. the set S, must be found and the dependencies between these skills, i.e. the surmise function σ, must be formulated. Second, questions must be constructed which can be used for diagnosis of the skills possessed by students. Then their connection to the skills, i.e. the skill assignment α, must be formulated. Third teaching materials, i.e. the set I, must be developed and it must be stated which instructions should be presented to a student in a special state of the student model by constructing the function δ.

These steps should be left to experienced teachers in the domain. But even if we assume that a group of such experienced teachers enforced such a necessary analysis of the domain we have to be aware of the risk that some of the structures σ, α, δ may contain errors which may influence the performance of the system negatively.

For example an indadequate construction of σ may lead to a student model $\mathcal{S}$ containing many states which will result in a very slow knowledge diagnosis.

This risk can be reduced with two different approaches. One approach is to test the assumptions contained in σ, α, δ empirically. How this can be done is for example described in Albert and Held (1994), Albert, Schrepp and Held (1994), Schrepp (1993), Held (1993), or Korossy (1993, 1997). But if the number of skills in S or the number of questions in Q is high this method may be problematic, since to many empirical data will be necessary to draw conclusions on the correctness of σ, α and δ.

A second approach to reduce the influence of an incorrect formulation of the basic structures is to make the functions σ, α, δ adaptive. We will illustrate this idea by an example. Assume that we have $I_j \in \delta(s, S')$, i.e. the experts assumed that I_j should be presented to a student mastering all skills from $S' \setminus \{s\}$ and not mastering s. Assume further that the instructional sequence I_j has not the intended effect, i.e. only a few students show mastery of s after I_j was presented to them. Remember that this can be recognized by the TEST mode. We can made

[4] This is realized within the system since the teaching procedure makes use of the surmise function σ to chose the skills which should be teached next.

δ easily adaptive if we assume that I_j is eliminated if less than a percentage p of students reaches mastery of s after I_j is presented to them. We have to ensure here that $\delta(s, S')$ will not become empty. Another possibility is to chose the instructional sequences accordingly to the probability of their success.

This example shows that it is relatively simple to make the system adaptive concerning δ, since we have a possibility to measure the success of an instructional sequence directly. For σ and α this is more complicated, because the correctness of these functions can not be observed directly. If, for example, a hypothetical skill state $S' \in \mathcal{S}$ is never observed by any student this may be due to the fact that $\mathcal{S}$ contains to much states, i.e. σ is formulated not strictly enough, or due to the fact that Q is not constructed adequately, i.e. there is no combination of solved and unsolved questions from Q which implies state S'. It is at the moment not clear how σ and α can be changed adaptively due to their success to explain the behavior of students.

We have formulated in this paper the structure of an ITS based on the theory of knowledge spaces and skill assignments. The next step in the development of our work should be the implementation of the ITS and its application in a special knowledge domain. Since we have described the components of our system as well as their interaction strictly formalized, the implementation of the system in a language of logical programming such as Prolog or Lisp should cause no problems.

A concrete application of this ITS can be used to evaluate the systems efficiency and can give raise to new insights in the dynamic of the teaching process, which may lead to further improvement of the theoretical structures.

ACKNOWLEDGEMENTS

The research reported in this paper was supported by Grant Lu 385/1 of the Deutsche Forschungsgemeinschaft to J. Lukas and D. Albert at the University of Heidelberg.

REFERENCES

Albert, D., & Held, T. (1994). Establishing knowledge spaces by systematical problem construction. In D. Albert (Ed.), *Knowledge structures* (pp. 81–115). Berlin, Heidelberg: Springer.

Albert, D., Schrepp, M., & Held, T. (1994). Construction of Knowledge Spaces for Problem Solving in Chess. In: G. Fischer & D. Laming (Eds.), *Contributions to Mathematical Psychology, Psychometrics, and Methodology.* New York: Springer.

Brown, J.S., & Burton, R.R. (1978). *Diagnostic models for procedural bugs in basic mathematical skills. Cognitive Science,* 2, 155-192.

Doignon, J.-P. (1994a). Knowledge spaces and skill assignments. In G. Fischer & D. Laming (Eds.), *Contributions to mathematical psychology, psychometrics, and methodology* (pp 111 -121). New York: Springer.

Doignon, J.-P. (1994b). Probabilistic assessment of knowledge. In D. Albert (Ed.), *Knowledge structures* (pp. 1–57). Berlin: Springer.

Doignon, J.-P., & Falmagne, J.-C. (1985). Spaces for the assessment of knowledge. *International Journal of Man-Machine Studies, 23*, 175–196.

Doignon, J.-P., & Falmagne, J.-C. (1998). *Knowledge spaces.* Berlin: Springer.

Falmagne, J.-C., & Doignon, J.-P. (1988). A markovian procedure for assessing the state of a system. *Journal of Mathematical Psychology, 32*(3), 232–258.

Held, T. (1993). *Establishment and empirical validation of problem structures based on domain specific skills and textual properties. A contribution to the theory of knowledge spaces.* Unpublished PhD thesis, University of Heidelberg.

Heines, J.M., & O'Shea, T. (1985). The design of a rule based CAI tutorial. *International Journal of Man–Machine Studies, 23*, 1–25.

Kimball, R. (1982). A self-improving tutor for symbolic integration. In: D. Sleeman & J.S. Brown (Eds.), *Intelligent tutoring systems.* London: Academic Press.

Kondo, H., Watanabe, K., Takeuchi, A., & Otsuki, S. (1990). ITS: Recognizing the context and guiding reduction processes for fraction calculation. *Proceedings of the International Conference on ARCE*, (pp. 31–37).

Korossy, K. (1993). *Modellierung von Wissen als Kompetenz und Performanz.* [Modeling of knowledge as competence and performance.] PhD thesis, University of Heidelberg.

Korossy, K. (1997). Extending the theory of knowledge spaces: A competence-performance approach. *Zeitschrift fr Psychologie, 205*, 53–82.

Kunichika, H., Takeuchi, A., & Otsuki, S. (1994). Hypermedia English learning environment based on language understanding and error origin identification. *IEICE Transactions on Information and Systems, E77-D (1)*, 89–97.

Lukas, J., & Albert, D. (1993). *Knowledge assessment based on skill assignment and psychological task analysis.* In G. Strube & K.F. Wender (Eds.), The cognitive psychology of knowledge: The German *Wissenspsychologie* project. Amsterdam: Elsevier (North–Holland).

Mandl, H., & Hron, A. (1985). *Förderung kognitiver Fähigkeiten und des Wissenserwerbs durch computerunterstütztes Lernen.* [Support of cognitive abilities and knowledge aquisition through computer–aided learning.] Report, Deutsches Institut für Fernstudien an der Universität Tübingen.

Matsuda, N., & Okamoto, T. (1992). Student model and its recognition by hypothesis-based reasoning in ITS. *IEICE Transactions on Information and Systems, J75-A*, 266-274.

Okamoto T., & Matsuda N. (1993). *Overview on the studies of intelligent CAIs/ITSs in Japan.* Unpublished report, Univ. of Electro–Communications Tokyo / Kanazawa Institute of Technology.

Ploetzner, R., & Spada, H. (1993). *Multiple mental representations of informations in physics problem solving.* In G. Strube & K.F. Wender (Eds.), The cognitive psychology of knowledge: The German *Wissenspsychologie* project. Amsterdam: Elsevier (North–Holland).

Schrepp, M. (1993). *Über die Beziehung zwischen kognitiven Prozessen und Wissensräumen beim Problemlösen.* [On the relation between cognitive processes and knowledge spaces in problem solving.] Doctoral dissertation, University of Heidelberg.

Sleeman, D., & Brown, J.S. (1982). *Introduction: Intelligent tutoring systems.* In: D. Sleeman & J.S. Brown (Eds.), Intelligent tutoring systems. London : Academic press.

Takeuchi, A., & Otsuki, S. (1994). An interactive learning environment for an intelligent tutoring system. *IEICE Transactions on Information and Systems, E77-D (1)*, 129–137.

Yasuda, K., & Okamoto, T. (1991). *The study of ITS to diagnose the algorithm of the novice's program.* ICOMMET'91. (Int. Conf. on Multi–Media in Education and Training)

9

Application of Doignon and Falmagne's Theory of Knowledge Structures to the Assessment of Motor Learning Processes

Susanne Narciss
Technische Universität Dresden

Investigating training methods such as mental rehearsal requires a detailed assessment of motor learning processes. Due to the lack of methods for investigating cognitive aspects of motor learning, the objective is to develop and validate research methods appropriate for assessing data on the internal representation and production of a motor action. On the basis of the theory of knowledge spaces introduced by Doignon and Falmagne (1985, 1998), we propose an enhanced application of this theory to develop these research methods: In order to establish theoretically based structures on the set of basic skills assumed necessary to perform the complex motor action breast stroke, we applied principles of cognitive and biomechanical task analysis. On the basis of this hypothetical skill structure we constructed cognitive tasks on the coordination of arm and leg movements in breast stroke. Thus it is possible to assess data on the internal representation and realization of breast stroke. The results of a motor learning experiment with 31 physical education students show that the developed research methods are appropriate for investigating the relation between the internal representation and the realization of a complex motor action.

INTRODUCTION

Supposing that motor learning consists of modifications of motor behavior and motor memory structures, the question arises of how to assess the motor and cognitive aspects of a motor learning process. This question is essential to the evaluation of training techniques such as mental rehearsal or imagery. Due to the lack of empirical procedures necessary to assess the modifications of motor memory structures, the evaluation of training techniques is generally implemented on the basis of data on motor behavior, whereas the cognitive aspects are neglected. Consequently, even the well-known efficiency of mental rehearsal is often underestimated and cannot be explained. The experimental objective of the work presented in this chapter was, therefore, to develop and validate procedures appropriate to assess data on motor and cognitive aspects of motor learning.

In 1985, Doignon and Falmagne presented a knowledge assessment procedure based on a theory of knowledge structures (Doignon & Falmagne, 1985). This theory is an approach to knowledge representation using set theory. The basic concepts such as knowledge state, knowledge structure, knowledge space and others are defined in terms of observable data structures. As this approach is not restricted to a particular domain of knowledge, we may use it as a basis for the development of the above-mentioned assessment procedures.

The purpose of this chapter is primarily to report the essential aspects of this research. In the first section, important results of a review of the motor learning research are summarized in form of a heuristic model of motor control and learning. Next, we propose an enhancement of Doignon and Falmagne's "theory of knowledge structures" which will be the basis for the development of procedures for the assessment of motor and cognitive aspects of motor learning. Then the application of this enhanced approach to the investigation of the learning of a particular complex motor action will be demonstrated in the next section. The last section summarizes a motor learning experiment, the purpose of which was to evaluate the applicability of the developed procedures.

MOTOR CONTROL AND LEARNING

One central goal of psychological approaches to the acquisition and/or the control of complex motor behavior is to explain how the motor system is controlled to produce coordinated and skilled actions. This motor system has high numbers of degrees of freedom. The plasticity and versatility of the motor control system thus requires the consideration of the following phenomena (e.g. Stelmach & Diggles, 1982; Zimmer & Koerndle, 1988):

- Complexity of the motor system: Bernstein (1967) identified the computational complexity of controlling many degrees of freedom. He counted

127 degrees of freedom, which means 127 independent coordinates of the motor system, that result in at least 2^{127} possible combinations. Effective motor control within this framework thus consists in mastering redundant degrees of freedom.

- Motor equivalence: Due to the complexity of the motor system, the same or equivalent motor actions may be produced by the use of several different muscle combinations. The functional role of muscles may vary from movement to movement (Hebb, 1949; Lashley, 1938).
- Motor variability: Even if identical environmental conditions are given, it is impossible to produce movement outcomes that are exactly identical in all kinematic and electromyographic details (e.g. Glencross, 1980; Schmidt, 1975).
- Motor flexibility: As a consequence of the nonspecificity of motor commands, we have to consider the enormous adaptability of the motor system. Experimental results show that even unexpected modifications of environmental conditions are compensated almost immediately (e.g. Kelso, Tuller, Vatikiotis-Bateson, & Fowler, 1984; Munhall & Kelso, 1985).

Approaches to motor control or learning should be evaluated regarding the ability to explain these issues. In the last decade there has been a controversial debate between motor and action approaches (e.g. Meijer & Roth, 1988): motor approaches try to explain the control of the degrees of freedom assuming an internal representation or centrally stored movement commands such as "memory traces" (Adams, 1971), "motor programs" (Keele, 1968), "generalized motor programs" (Schmidt, 1975), and so on. They attribute considerable responsibility to the program that defines all aspects of the intended movement in advance. Thus, motor approaches cannot satisfactorily address the issues of motor equivalence, motor variability and motor flexibility. Action approaches, on the other hand, attempt to delineate the production of coordinated motor actions by analyzing the interactions between the motor system and its environment. Based on Gibson's ecological theory of direct perception (Gibson, 1979), they deny the existence of internal representations and look for fundamental organizing principles producing order and regularity in motor actions (e.g. Reed, 1982; Turvey, 1977). Ecological models of motor control reduce the complexity of the motor system resulting in a loss of accuracy. To a certain extent, motor variability, and motor equivalence may be explained. Action approaches, however, fail to explain the enormous flexibility of the human motor system. They are too restricted to model the acquisition of a motor skill or the qualitative characteristics of a skilled versus an unskilled production of a motor action (see also, Zimmer & Koerndle, 1994).

Within the debate between these two approaches, the concept of internal representation has undergone a major metamorphosis. It has evolved from a program-like internal structure defining all movement commands in advance to a loosely defined plan of action. In fact, even strong defenders of action approaches agree

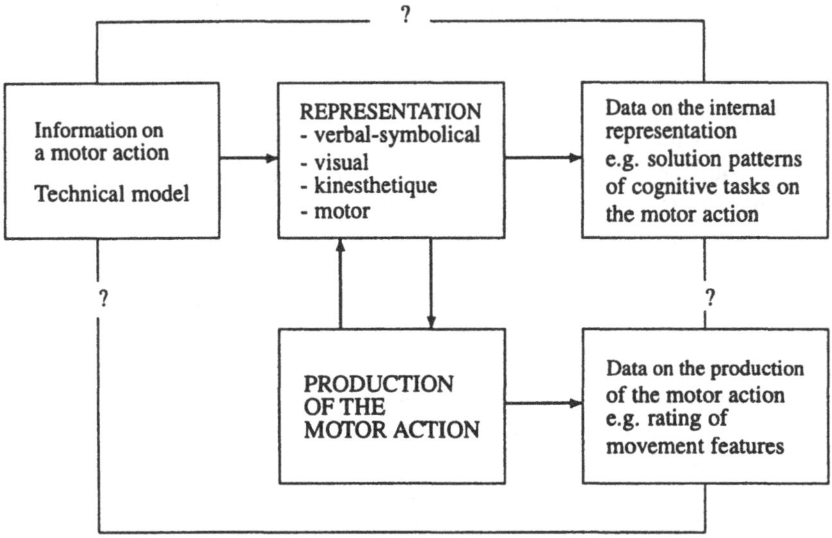

FIG. 1. Heuristic model of motor control and learning (Narciss, 1993)

with the assumption that the acquisition of a new state of a complex system is specified from within the system (e.g. Turvey, 1994). A review of motor learning research offers empirical support for the idea that at least essential features and organizing principles of a motor action are represented internally. It is almost impossible to explain positive effects of training techniques such as mental rehearsal or imagery, on the basis of pure ecological theories. Considering the metamorphosis of the concept of internal representation of a motor action, we use this concept to describe memory structures necessary to solve motor and cognitive tasks for a given motor action. Using the term *representation*, we want to emphasize that systematic relations between the structure of the given motor action (defined by its essential features and organizational principles, fixing the possible relations between these features) and the corresponding internal structure exist, supposing however, that this internal structure does not determine the motor action in the sense of a motor program.

With regard to this view of the concept of internal representation of a motor action, a heuristic model of motor control and learning is presented that serves, on the one hand, to analyze and structure the results of motor learning research systematically, and, on the other hand, to derive strategies for how to search for characteristics and/or modifications of the internal representation of a motor action (see Fig. 1).

Our heuristic model of motor control and learning is based on the assumption that learning a complex motor action requires the internal representation of the essential features and organizational principles of this motor action. Information about these features and principles may be obtained externally by instruction or correction by teachers or coaches, by descriptions in books or other print-media, or by all forms of audiovisual media. Due to the complexity of the motor system, however, there is another important source of information in a motor learning process: The production of the motor action offers information about kinesthetique aspects of the motor action to be learned. Consequently, we may attribute a guidance function to the internal representation, although we have to take into account that not only externally presented information, but also each production of the motor action can result in modifications of the internal representation and that these modifications may have consequences for the next production of the motor action. To emphasize this interaction and the different sources of information, we suppose that there are at least four modalities to represent a motor action (e.g. Heuer, 1985; 1990):

- a motor modality representing efferent aspects of the motor action,
- a kinesthetique modality representing the afferent aspects emerging from the production of the motor action,
- a visual modality representing the visible aspects, and
- a verbal-symbolic modality representing conceptual aspects of the motor action.

Assuming that motor control consists in information processing principles as well as in self-organizing structures, motor learning may be defined as generating, differentiating and structuring of the multimodal internal representation, and the production of a motor action based on the processing of information obtained externally and internally (e.g. Heuer, 1983).

ASSESSMENT OF MOTOR LEARNING BASED ON THE THEORY OF KNOWLEDGE STRUCTURE

The thesis presented in this chapter deals mainly with the development and evaluation of procedures to assess modifications of the internal representation and of the production of a complex motor action, taking into account the characteristics of human movement mentioned above. Important steps in solving this task are (a) the analysis and formal description of the motor action under observation, (b) the definition of performance and knowledge structures of this motor action and (c) the construction of motor and cognitive tasks as instruments of the assessment procedure.

Background

Because the following formalizations are based on the enhanced set theory approach to knowledge representation we briefly outline those concepts necessary to understand our approach.

Basic Concepts

Within the framework of the theory of knowledge structures, a particular knowledge domain is conceptualized as a set Q of questions or problems. The set of questions a subject can answer with respect to this knowledge domain is called a *knowledge state* and is characterized as a subset of this set of questions: $K \subseteq Q$. Assuming that Q has specific structural properties, not every possible subset of Q describes an admissible knowledge state. An efficient knowledge assessment procedure with respect to the knowledge domain under consideration is thus not based on the family of possible knowledge states, the power-set of Q, but on a collection $\kappa \subseteq 2^Q$ of all admissible knowledge states. In order to determine the admissible knowledge states, Doignon and Falmagne (1985) propose either

- the *surmise-relation* S — a binary relation on the set of questions

  $$\kappa := \{K \subseteq Q \mid \text{if } q \in K \text{ and } pSq, \text{ then } p \in K \text{ for all } q, p \in Q\}$$

- the *surmise-system* σ which assigns to every element of Q a set of subsets of Q:

  $$\kappa_{(Q,\sigma)} := \{K \subseteq Q \mid \text{ for all } q \in K \text{ there exists a } C \in \sigma_{(q)} \text{ with } C \subseteq K\}$$

These two surmise concepts express formally that knowing the answer to a given question q implies the solving of some further problems related to this question. A *knowledge structure* (Q, κ) is thus defined by a nonempty set Q of questions and a collection of admissible knowledge states.

Skill assignment

A crucial problem in applying this theory of knowledge representation is the question of how to obtain a surmise relation or a surmise system. As already mentioned, we used procedures of skill assignment to solve this problem. These procedures are based on the classical idea that domain-specific knowledge may be explained by skills. A problem is solved if the subject possesses the required skills (see Marshall, 1981). Based on psychological task analysis, a list of skills (in the sense of elementary units of knowledge) has to be specified for every item in a set of problems, this list being necessary to solve the corresponding item. Doignon (1991) formalized the links between skills and problems as follows:

DEFINITION 9.1 Let Q be a set of problems. A skill assignment on Q is formed by a nonempty set S of skills, and a mapping τ from Q to the family of all nonempty subsets of S. This skill assignment is denoted as (Q, S, τ), and essentially consists of a mapping $\tau : Q \to 2^S \setminus \{\emptyset\}$. The skills belonging to $\tau(q)$ are said to be assigned to q. □

Information system (Scott, 1982)

The assessment of knowledge as presented by Doigon and Falmagne (1985) and Falmagne, Koppen, Villano, Doignon and Johannesen (1990) is based on a classification of problem solutions as being correct or incorrect. There are, however, many knowledge domains where a more detailed differentiation of problem solutions is desirable. It is especially important in the assessment of motor learning processes to analyze individual productions of a motor action in as much detail as possible considering the characteristics of human movement. Taking into account not only correct or incorrect, but also several specified categories of solutions, the enhancement of the theory of knowledge structures is necessary. Besides the well-known methods to establish the structure of a problem set (surmise-relation; surmise-system), we need a principle to define which categories of solutions are compatible or incompatible with others. Lukas (1997) used the algebraic structure of an information system introduced by Scott (1982) to formalize this enhancement:

DEFINITION 9.2 An information system is a triple $S = \langle S, C, \vdash \rangle$ such that for all $A, B, C \subseteq S$ the following properties hold:

- S is a set of "units of information"
- C is a non-empty set of finite subsets of S (the finite consistent sets), with the following properties:
  INF 1: $A \in C$ and $B \subseteq A$ implies $B \in C$,
  INF 2: $a \in S$ implies $\{a\} \in C$
- $\vdash$ is a relation from C to $S (\vdash \subseteq C \times S)$ with
  INF 3: $A \in C, a \in S$ and $A \vdash a$ implies $A \cup \{a\} \in C$,
  INF 4: $A \in C$ and $a \in A$ implies $A \vdash a$
  INF 5: If $A, B \in C$ and $a \in S$ are such that $A \vdash b$ for all $b \in B$ and $B \vdash a$ then $A \vdash a$. □

This algebraic structure axiomatizes on the one hand the idea that some knowledge implies some further knowledge (see the surmise-concepts) and on the other hand, it formalizes the possible inconsistencies of certain skill-combinations (see Lukas, 1997).

Modelling central aspects of a complex motor action

In this section, the concept of "complex motor action" is defined and formal concepts are specified to describe the internal representation and the production of a motor action.

Formal Description of a Complex Motor Action

Biomechanical analysis of complex motor actions show that human motor actions are characterized by patterns of coordinated movement features. "Walking" for example may be described by the features for each leg (e.g. standing on the leg, pulling, swinging, landing) and by the structural rules that coordinate the two legs (e.g. when standing on the right leg, the left leg is swinging forward). Using concepts of set theory, we describe a complex motor action by sets of movement features and their possible combinations. To determine the possible combinations of movement features, we need at least two principles:

- one principle that serves to define combinations of features implied by the combination of some other features, and
- a second principle that formalizes the compatibility of features (e.g. it is impossible to stand on one leg and to swing this leg at the same time).

Considering the algebraic structure presented above, a complex motor action may thus be characterized in terms of an information system:

A complex motor action M is defined as a triple $M = (S, C, \models)$. S describes the Cartesian product of all the feature sets characterizing the motor action. C is a collection of finite consistent subsets of S determining the compatibility of feature combinations. $\models$ is a relation that, in the case of a simple motor action, is equivalent to the relation $\vdash$ of an information system. For rather complex motor actions, we do however, interpret the relation $\models$ as a surmise-mapping. A subset of S is called an element of M if it is contained in C and closed with respect to the relation $\models$.

Sensorimotor Performance Structure

To establish economical procedures for the assessment of a subject's individual production of the motor action under analysis, we have to define the sensorimotor performance domain in terms of observable data-structures. On the basis of a descriptive technical model of the motor action standardizing the optimal quality for each movement feature, different grades of quality are assigned to these features:

DEFINITION 9.3 Let us consider a complex motor action M characterized by M_j sets of movement-features. For each $m \in M_j$ a set of possible grades of quality $A_{mj} = \{m_{j^0}, m_{j^1}, m_{j^2}, ...m_{j^n},\}$ is specified. m_{j^0} describes the worst grade of quality, m_{j^n} the best grade. The sensorimotor performance domain is denoted as the union of all possible variants of producing the motor action M:

$$P_M := \cup A_{m_j} \setminus m_{j^0} . \qquad \square$$

A subject's individual performance state may thus be defined as a subset of the sensorimotor performance domain:

$$PS \subseteq P_M .$$

Introducing different grades of quality for each movement feature of a complex motor action, we have to take into account many more performance states as if we were classifying the different variants of producing a given motor action only as either correct or incorrect. As there are logical implications or incompatibilities for the possible admissible combinations of these movement features, we apply the principles axiomatized by the algebraic structure of an information system to define the set of admissible sensorimotor performance states:

$$R_{mot} \subseteq 2^{P_M}.$$

The possible inconsistencies of certain combinations of movement features are formalized as follows:

$$C := \{PS \subseteq P_M \mid \forall m_j \in M_j; m_{j^x} \in PS \Rightarrow m_{j^y} \notin PS, \forall x \neq y\}$$

Picking up the idea that the production of specific grades and combinations of some movement features implies the quality of some further movement features, we use the relation in the sense of a surmise-mapping. This relation is derived from the biomechanical assumptions described by the technical model of the motor action under consideration. A performance state PS is called an element of R_{mot} if it is contained in C and closed with respect to the relation $\models$.

A sensorimotor performance structure (P_M, R_{mot}) thus consists of a union of possible variants of producing a complex motor action, called sensorimotor performance domain P_M, together with a collection R_{mot} of admissible sensorimotor performance states.

Sensorimotor Knowledge Structure

If we consider the question of how to assess modifications of the production and of the internal representation of a complex motor action, it is necessary to describe the internal representation of a complex motor action in terms precisely related to those concepts defining a subject's individual production of this motor action.

Taking into account the formal description of the motor action under investigation, elementary sensorimotor skills s_j are specified on the basis of a psychological task analysis. By analogy with the establishment of the performance domain, different grades of quality are assigned to each of these elementary sensorimotor skills. The sensorimotor knowledge-domain is denoted as the union of all possible varieties of quality of these elementary skills:

$$K_M := \cup A_{s_j} \setminus s_{j^0}.$$

Similar to the definition of a sensorimotor performance state, we may describe a sensorimotor knowledge state KS by a subset of the sensorimotor knowledge-domain:

$$KS \subseteq K_M.$$

Because we are considering different grades of quality of the elementary sensorimotor skills, we have to exclude possible inconsistent combinations of skills (e.g. the combination of different grades of a specific skill):

$$C := \{KS \subseteq K_M \mid \forall s_j \in S_j; s_{j^x} \in KS \Rightarrow s_{j^y} \notin KS, \forall x \neq y\}.$$

Assuming—as we did in our heuristic model—that interaction exists between the internal representation and the production of a complex motor action, we infer the rules determining which combinations of sensorimotor skills imply further skills or further combinations of skills from those implication-rules used to establish the performance structure. That means that, besides the specification of inconsistent skill-combinations, we apply the relation $\models$ to determine a set of admissible knowledge states:

$$R_{cog} \subseteq 2^{K_M}.$$

In order to ascertain a subject's sensorimotor knowledge state empirically, a set T_{cog} of cognitive tasks concerning the complex motor action under consideration must be constructed. With respect to the results of the psychological task analysis of the motor action, each task is designed using the procedure of a skill assignment. By a mapping $\mu : K_M \rightarrow 2^{R_{cog}}$ each task $t \in T_{cog}$ is associated with a set $\mu(t)$ of sensorimotor knowledge states. From the solution of these tasks we may infer a subject's knowledge state.

PERFORMANCE AND KNOWLEDGE STRUCTURES FOR THE COMPLEX MOTOR ACTION "BREAST STROKE"

The following demonstration intends to illustrate the ideas formalized above. It shows how to establish performance and knowledge structures and how to construct cognitive tasks with respect to the particular motor action, "breast stroke".

Task analysis of breast stroke

The breast stroke is characterized by propeller-like symmetric arm and leg sweeps. In contrast to other swimming styles (crawl, butterfly, etc.) the recovery of arms and legs has to be carried out under the surface of the water. The coordination of the arm and leg sweeps is essential for effective propulsion. Wrong timing of the arm and leg sweeps causes drag that normally results in a reduction in speed.

Taking the results of biomechanical analyses of the breast stroke into account (see, e.g., Chollet, 1990; Maglischo, 1982; Reischle, 1988) a swimming cycle

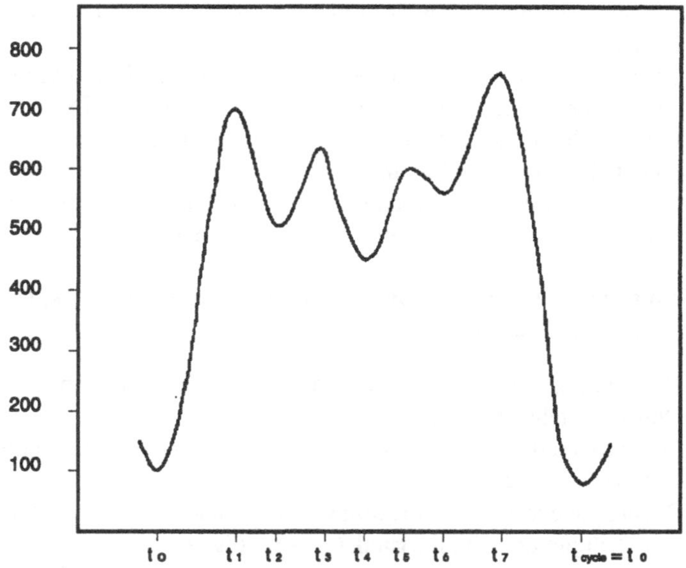

FIG. 2. Intracyclical velocity plot of breast stroke swimming

may be divided in four propulsive phases and the overlapping recovery of arms and legs (see the velocity plot in Fig. 2):

- The first propulsive phase is characterized by the outsweep of the legs: The circling outward and downward is started from a position where the feet are near the buttocks by dorsiflexing and pronating the feet as much as possible ($t_0 - t_1$).
- The second propulsion results from the insweep of the legs, which begins when the legs are nearly extended. This terminates the downsweep. The feet are inverted and pitched inward as much as possible and the legs sweep powerfully inward across the water until they come together ($t_1 - t_3$).
- A third propulsive force is provided by the catch following the outsweep of the arms: The hands pitched outward, downward, and backward sweep outward and downward in a circular path until they approach their deepest point. The speed of the hands should be accelerated during the sweep ($t_3 - t_5$).
- The fourth propulsion is caused by the insweep of the arms: After a circular transition from downsweep to insweep (see t_4 of the velocity plot, Fig. 2), the hands sweep inward, upward, and backward. The movement inward and upward of the hands should be accelerated throughout the insweep ($t_5 - t_7$).
- The recovery of the arms consists of a gentle slip forward of the hands and a stretching motion of the whole arms related to the outsweep. The purpose

of this stretching motion is to place the arms in position for an effective catch ($t_7 - t_0$).

- The recovery of the feet starts after a definite glide of the legs just when the upward force of the arms causes the hips to be lowered. The legs, which are kept together in horizontal alignment with the trunk while the arm stroke is executed, relax at the hip and knee joints. The heels are rapidly and gently brought upward and forward toward the buttocks by flexing the knees, followed by some flexion and outward rotation at the hip joints ($t_7 - t_0$).

To obtain an optimal coordination of the arm stroke, and leg kick, the following rules should be considered:

- The circling outward and downward of the legs starts when the arms are almost stretched (t_0 , Fig. 2).
- The outsweep of the arms begins when the legs are terminating their insweep (t_4, Fig. 2).
- During the execution of the arm stroke the legs are kept together in a horizontal alignment with the trunk ($t_4 - t_6$, Fig. 2).
- The legs are flexed at the knee joints when the recovery of the arms begins (t_7 , Fig. 2).

Determination of performance and knowledge states on the basis of the psychological task analysis

To specify basic movement-features necessary to perform the breast stroke in an optimal way, we have to extract those units of action that are supposed to be controllable during swimming. With respect to the schema-theoric approach presented by Zimmer and Koerndle (1988), this specification has to be done taking into account the performance level of the subjects-in our case, students of physical education.

In order to describe the *arm stroke (A)*, at the very least, we consider the following basic elements:

- propulsive sweeping outward and downward (A_o)
- propulsive sweeping inward and upward (A_i)
- gently slipping the hands forward (A_s)
- stretching, almost gliding motion of the arms (A_g)

The *leg kick (L)* consists of the elements:

- propulsive sweeping outward and downward (L_o)
- propulsive sweeping inward (L_i)
- keeping the legs together and stretched, glide (L_g)
- gently recovering the heels (L_r)

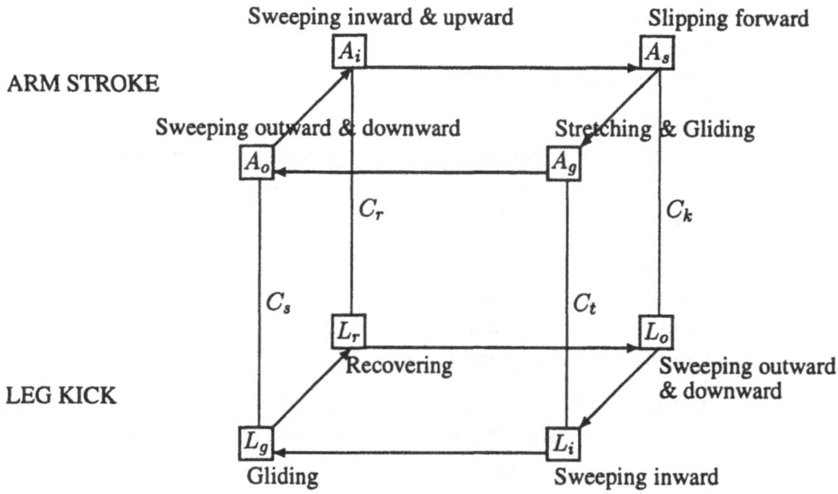

FIG. 3. Structure of breast stroke

The links between these basic elements could be derived from the rules determining an optimal *coordination (C) of the arm stroke and leg kick.* These rules are denoted as follows (see Fig. 3):

- C_k The kick of the legs starts when the arms are almost stretched.
- C_s The arm stroke begins at the end of the insweep of the legs.
- C_r The recovery of the legs begins at the end of the insweep of the arms (definite glide of the legs).
- C_t The transition from the propulsive phase of the leg motion to the propulsive phase of the arm motion is characterized by a definite stretching forward of the arms.

In order to determine the sensorimotor performance domain we specify for each movement-element of the arm stroke and leg kick two grades of quality (correct or incorrect), whereas we attribute three grades of quality (optimal = 2; with little deviation = 1; completely incorrect = 0) to the features associated with the coordination rules. Thus performance domain of breast stroke (P_B) may been denoted as

$$P_B = \{a_o, a_i, a_s, a_g, l_o, l_i, l_g, l_r, c_k^2, c_k^1, c_s^2, c_s^1, c_r^2, c_r^1, c_t^2, c_t^1\}$$

The possible implications and inconsistencies of element combinations are derived from the movement structure illustrated in Fig. 3:

1. Assuming that the optimal production of the coordination features is based on the correct production of the adjoining arm and leg features we formulate:

c_s^2 implies a_o and l_g
c_k^2 implies a_s and l_o
c_r^2 implies a_i and l_r
c_t^2 implies a_g and l_i

2. As swimming is a cyclical motor action, it is plausible to suppose that if a coordination feature is produced in an optimal way the adjoining coordination features are not completely incorrect:

   c_s^2 implies (c_r^2 or c_r^1) and (c_t^2 or c_t^1)
   c_k^2 implies (c_r^2 or c_r^1) and (c_t^2 or c_t^1)
   c_r^2 implies (c_s^2 or c_s^1) and (c_k^2 or c_k^1)
   c_t^2 implies (c_s^2 or c_s^1) and (c_k^2 or c_k^1)

3. Considering three grades of quality for the coordination features, we take the following inconsistencies of feature combinations into account:

   c_s^2 is inconsistent with c_s^1
   c_k^2 is inconsistent with c_k^1
   c_r^2 is inconsistent with c_r^1
   c_t^2 is inconsistent with c_t^1

Applying algorithms developed by Lukas and Unnewehr (see Unnewehr, 1992) on the basis of the algebraic structure of an information system, the set of possible performance states ($2^{16} = 65536$) was reduced to a set $\mathbf{R}_{\mathbf{mot}^\mathrm{B}}$ of 2720 admissible performance states.

On the basis of the task analysis summarized above, we may now specify the sensorimotor knowledge domain of the breast stroke. Referring to the heuristic model of motor control and learning we consider that

- knowing the basic elements of the arm stroke ($k_{A_o}, k_{A_i}, k_{A_s}, k_{A_g}$),
- knowing the basic elements of the leg kick ($k_{L_o}, k_{L_i}, k_{L_g}, k_{L_r}$) and
- knowing the basic coordination rules ($k_{C_r}, k_{C_t}, k_{C_s}, k_{C_k}$)

is essential for an optimal production of the breast stroke. To guarantee a precise relation between performance and knowledge domains we attribute two grades of quality to those cognitive skills associated to the basic movement features for the arm stroke and leg kick and three grades of quality to those assigned to the rules determinating the coordination of arm and leg motions. The sensorimotor knowledge-domain of breast stroke is thus defined as:

$$K_B = \{k_{A_o}, k_{A_1}, k_{A_s}, k_{A_g}, k_{L_o}, k_{L_i}, k_{L_g}, k_{L_r}, k^2_{C_r}, k^1_{C_r}, k^2_{C_t}, k^1_{C_t}, k^2_{C_s}, k^1_{C_s}, k^2_{C_k}, k^1_{C_k}\}$$

Corresponding to the performance domain, the possible combinations of these basic skills were reduced, respecting the implications and inconsistencies specified above. The set $\mathbf{R}_{\mathbf{cog}^\mathrm{B}}$ thus consists of 2720 admissible knowledge states.

Constructing cognitive tasks on the coordination of breast stroke

Finally, taking the task analysis of breast stroke and the set of admissible knowledge states into account, we could design cognitive tasks appropriate to ascertain a subject's knowledge state. Because a correct coordination of the arm stroke and leg kick implies knowing the links between their basic elements, we decided to construct rather complex cognitive tasks that require subjects to associate pictures of several arm stroke elements with pictures of particular leg kick elements (see Fig. 4). These pictures were produced taking video prints from those positions of a swimming cycle linked to a minimum or a maximum of the velocity plot presented in Fig. 2. After dividing each video print into two parts-a part showing the features of the arm stroke, the other part those of the leg kick-these parts were drawn separately. Each of the basic knowledge elements of the arm stroke and the leg kick could be precisely assigned to two pictures. The elements representing the basic coordination rules were associated with a specific order of these pictures (see Table 1). On the basis of this assignment of the sensorimotor knowledge elements, we constructed three problems on the coordination of breast stroke: The first problem asked the subjects to assign the pictures of the leg kick to those of the arm stroke (see Fig. 4), the second, to find the sequential order of the arm stroke pictures, and the third, to find the sequential order of the the pictures of the leg kick. Subjects had to do multiassignments of those pictures, associated with the basic elements "stretching-almost gliding motion of the arms" (A_g) and "keeping the legs together and stretched-glide" (B_g). These multiassignments were included to avoid the application of algorithms, such as finding the sequential order of the arm stroke and the leg kick, thereby solving the structuring problem.

In order to determine a subject's sensorimotor knowledge state, the solutions of these three cognitive tasks were analyzed with regard to the basic elements of the knowledge domain. We assume that, for instance, a correct solution of the first task indicates the application of the knowledge elements of coordination. The observed solution patterns are compared with the admissible knowledge states. If there are deviations we look for the symmetric distances between solution patterns and the closest admissible knowledge state.

EXPERIMENTAL RESULTS

After initial empirical validations of the cognitive tasks on breast stroke that show that these tasks are appropriate to assess data on the internal representation of this complex motor action, we applied the assessment procedure in a motor learning experiment. In a quasi-experimental study, we compared the modifications of the internal representation and those of the production of the breast stroke of two experimental groups participating in a 6-week training period. The purpose of this quasi-experimental study was to investigate if a training program results

Problem 1:

On the following pages you will find several pictures of the arm stroke and the leg kick of breast stroke. These pictures represent basic elements of a breast stroke cycle.

- Please assign the pictures of the leg kick to the corresponding pictures of the arm stroke. You may solve this problem by multiassignment of some pictures. That means that there may be more than one picture of the leg kick corresponding to a particular picture of the arm stroke and vice versa.
- Please rate for each assignment, how sure you are, that it is correct:
  1 = I am really sure that the assignment is correct
  6 = I don't know if the assignment is correct

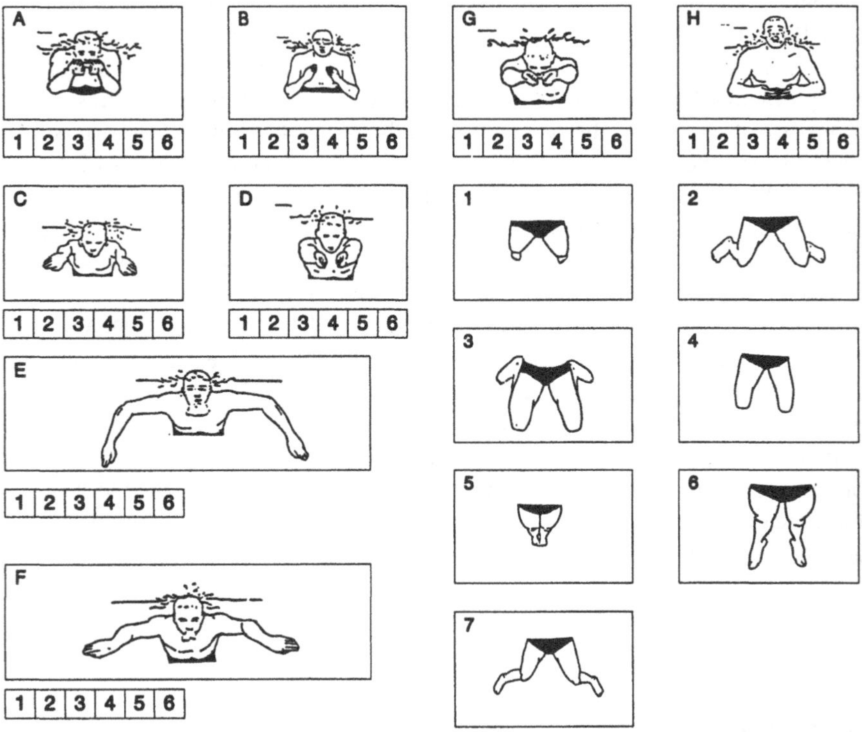

FIG. 4. Cognitive task on the coordination of breast stroke

TABLE 1
Assignment of the Basic Knowledge Elements
to the Components of the Cognitive Task on Breast Stroke

Basic Knowledge Elements	*Task Components*
Armstroke	**CFEBAGGD**
propulsive sweeping outward & downward (A_o)	C F
propulsive sweeping inward and upward (A_i)	E H
gently slipping forward the hands (A_s)	B A
stretching - gliding motion of the arms (A_g)	G G D
Legkick	**327655514**
propulsive sweeping outward & downward (L_o)	3 2
propulsive sweeping inward (L_i)	7 6
keeping legs together and stretched - glide (L_g)	5 5 5
gently recovering the heels (L_r)	1 4
Coordination	
start of the leg kick (C_k)	C-5
start of the arm stroke (C_s)	A-3
begin recovery of the legs (C_r)	H-1
transition leg kick/arm stroke (C_t)	D-6

in different modifications of the internal representation and of the production of the breast stroke than the usual training program if this consists of a systematic combination of mental rehearsal and physical exercises.

Method

Although previous research on motor learning neglects data on the internal representation of the motor action to be learned, the thesis presented in this chapter attempts to assess such data using the example of the production of breast stroke.

Subjects

The quasi-experimental study was conducted with two swimming classes in the physical education department at the university of Heidelberg. The students of class A were assigned to a mental rehearsal treatment, the students of class B attended the usual swimming class. Thirty-one physical education students, 17 of the mental-rehearsal group (mean age 24.5), and 14 of the control group (mean age 23.9), participated regularly in the motor learning experiment.

Material

In order to ascertain data on the production of the breast stroke, the subjects were asked to swim 25 meters. Each subject's movement was recorded frontally by an underwater video camera. The cyclic variations of velocity were registered with the velocity sensor GS9 (ASM, München-Unterhaching; for a detailed description see Reischle & Spikermann, 1992). The videotapes were analyzed using an observation instrument that was constructed on the basis of the task analysis mentioned above. Two independent swimming experts had to rate the grade of quality for each basic element of the performance domain. The plots of the intercyclical variations of velocity helped the raters to score the items in case of doubt. By analogy to the determination of a subject's knowledge state, the observed movement patterns are compared with the admissible performance states. In case of deviations, the symmetric distances between solution patterns and the closest admissible performance state is calculated.

The data on the internal representation of the breast stroke was obtained by asking the subjects to solve the cognitive tasks presented in the previous section. The cognitive data set was analyzed with respect to the knowledge domain (see previous section).

Besides the determination of knowledge or performance states on the basis of the theory of knowledge-structures, the research question mentioned above requires the assessment of modifications of the quality of the determined knowledge of performance states. Because the given knowledge and performance domain are too complex to be represented in a Hasse-diagram, the grades of quality linked to the basic elements of these domains were scored as follows:

- each correct solution assigned to a basic element of the arm stroke
  → 1 point
- each correct solution assigned to a basic element of the leg kick
  → 1 point
- each correct solution assigned to a coordination element
  → 2 points
- each solution with little deviation assigned to a coordination element
  → 1 point
- each incorrect solution of a basic element
  → 0 points

All item scores were added up to yield at the overall score of each knowledge or performance state (from 0 to 16).

Procedure

Prior to the first training session and 1 week after the last training session of the 6-week training period, subjects of both experimental groups had to solve the cognitive tasks on breast stroke within their swimming lessons. The performance data were obtained 3 days later within the next swimming lesson. During the 6-week training period the subjects of the control group (N = 14) participated twice a week in the usual swimming classes. A specific mental training program was designed for the experimental group (N = 17) using the research results on mental rehearsal (for a review see, e.g., Feltz & Landers, 1983). At the beginning of each swimming lesson the subjects were asked to do particular mental exercises followed by the corresponding motor exercises. Subjects of the experimental group spent 15 minutes of the swimming lesson (45 minutes) rehearsing mentally particular aspects of the coordination of the breast stroke. The content of the mental exercises was chosen with regard to the specific errors of coordination. In the first few sessions, the objective was to reconstruct and reflect the individual representation of the breast stroke. Using videos of swimming experts, the subjects had to compare their individual representations of the breast stroke with the experts' swimming performances. The aim of the next few sessions was to differentiate and correct the individual representation. Subjects were asked to solve verbalization and visualization tasks, as well as tasks focusing on the sensoric aspects of correct coordination. In order to consolidate the corrections in the last term of the training period, the subjects were asked to observe the performances of other students, to detect movement errors, and to give instructions for their correction.

Results

Data analysis of the pretest and posttest performances in solving the motor and the cognitive tasks on the breast stroke provide some interesting information about the modifications of the data sets. Table 2 shows that significantly more admissible knowledge states could be observed for the experimental group in the posttest (14) than in the pretest (7), whereas there is little difference between the number of the observed knowledgestates of pre- (7) and posttest (9) of the control group. The symmetric distances of the observed non-admissible knowledge states and the closest admissible knowledge states were rather small (see Table 2) .

Mean scores of the pre- and posttest show that there was a homogenous inital level of the observed knowledge states for both experimental groups (mean score MT-group = 6,76; mean score control group = 6,85). Whereas the mental training group obtained significantly more scores for solving the cognitive tasks in the posttest (mean score 10,11), the scores of the control group did not increase significantly (mean posttest score 8,14). Error analysis of the observed solution patterns of the pretest show that there is an accumulation of wrong solutions for those task components assigned to the basic elements "stretching-almost gliding motion of the arms (A_g)" , "propulsive sweeping inward of the legs (L_i)", "glide of the legs (L_g)", and "gently recovering the heels (L_r)". Subjects of the MT group made significantly less errors when solving these task components in the posttest. There is, however, no comparable effect for the control group.

Table 3 shows no significant differences between the number of admissible

TABLE 2
Sensorimotor Knowledge States of the Pretest and the Posttest

Treatment	Test	Knowledge domain 2^{K_B}	R_{cog}	observed admissible knowledge states	observed non-admissible knowledge states	Frequency of distances between observed knowledge states and closest states of R_{cog} 0	1	2	3
MT	t1	65 536	2720	7	10	7	9	1	-
(N=17)	t2	65 536	2720	14*	3*	14	3	-	-
Control	t1	65 536	2720	7	7	7	7	-	-
(N=14)	t2	65 536	2720	9	5	9	4	1	-

2^{K_B} = set of possible skill combinations
R_{cog} = set of admissible skill combinations
MT = Mental Training

performance states in the pretest and posttest for both experimental groups. The symmetric distances of the observed non-admissible performance states and the closest admissible performance states were rather small (see Table 3) .

Mean scores of the pretest and posttest show that the performance level of both experimental groups improved significantly (mean score MT-group(t1) = 9,47; mean score MT-group (t2) = 13,00; mean score control group (t1) = 10,21; mean score control group (t2) = 12,85). Error analysis of the observed performance states of the pretest show that the movement components assigned to the basic elements "propulsive sweeping inward of the arms (A_i)" , "propulsive sweeping inward of the legs (L_i)", and "Glide of the legs (L_g)" were often increasingly incorrect. Whereas subjects of the MT group made significantly less errors in the posttest, when producing the movement components linked to A_i and to L_g, there is no comparable effect for the control group.

Comparing the results concerning the knowledge states with those concerning the performance states, the following aspects should be noted:

- For the subjects of the MT group, significantly more admissible knowledge states were observed in the posttest than in the pretest. At first glance, there is no comparable increase in admissible performance states. An analysis of the violations responsible for the deviations of the non-admissible performance states shows, however, that in most cases, the distance is due to errors linked to the basic element "propulsive sweeping inward and upward of the arms (A_i)". A correct production of this element requires a strong constitution of the swimmer. The analysis of both the non-admissible knowledge states and the non-admissible performance states of the control group

TABLE 3
Sensorimotor Performance States of the Pretest and the Post Test

Treatment	*Test*	*Perfomance domain* 2^{P_B}	R_{mot}	*observed admissible performance states*	*observed non-admissible performance states*	*Frequency of distances between observed performance states and closets states of* R_{mot} 0	1	2	3
MT	t1	65 536	2720	7	10	7	7	3	-
(N=17)	t2	65 536	2720	8	9	8	7	2	-
Control	t1	65 536	2720	6	8	6	7	-	1
(N=14)	t2	65 536	2720	8	6	8	4	1	1

2^{P_B} = set of possible combinations of movement features
R_{mot} = set of admissible performance states
MT = Mental Training

yielded no comparable peculiarities.

- When considering mean scores and standard deviations of the observed data structures, it should be emphasized that there is a significant group effect with respect to the different treatments. Mean scores of the MT group were significantly higher for both, the knowledge and the performance states. Although there is a signifcant improvement of the control group's performance states, we could not observe significant modifications from pre- to posttest in solving the cognitive tasks.
- A detailed error analysis reveals that, especially in the pretest, subjects made the same errors in solving the cognitive tasks and in producing the coordination of the breast stroke. As mentioned above, errors are increasingly observed for those task components linked to the basic elements "propulsive sweeping inward of the legs (L_i)" and "glide of the legs (L_g)". It is worth noting that most of the subjects of the MT group improved in solving these task components, whereas there are no comparable modifications within the data of the control group.

CONCLUSION

The above findings indicate that the application of the enhanced theory of knowledge structures offers possibilities to assess modifications of the internal representation and of the production of a complex motor action. In fact it is remarkable that even the calculation of the symmetric distances of the knowledge- and performance states throw some light on systematic modifications such as typical errors or corrections of specific elements. We have to emphasize, however, that due to the complexity of the motor action under investigation the problem arised how to determine the quality of the observed knowledge or performance states. We solved this problem using a simple rating strategy. Therefore, however, specific modifications where initially disregarded. Concerning the question of wether mental rehearsal has specific effects on the internal representation and on the production of a complex motor action, the consideration of these specific modifications, is of great importance. To account for these modifications the assessment procedure had to be completed by an error analysis. The results of this error analysis show some analogous modifications of the data on the internal representation and of the data on the movement production. Subjects' errors may thus be considered as in a detailed analysis of structural modifications during a motor learning process. Lukas (1997) proposed an enhancement of the theory of knowledge structures that allows taking typical errors of a given knowledge domain into account. As this enhancement is based on the algebraic structure of an information system, it may be added to the procedure presented in this chapter.

ACKNOWLEDGEMENTS

The research reported in this paper was supported by Grant Lu 385/1 of the Deutsche Forschungsgemeinschaft to J. Lukas and D. Albert at the University of Heidelberg.

REFERENCES

Adams, J. A. (1971). A closed-loop theory of motor learning. *Journal of Motor Behavior, 3*, 111–150.

Bernstein, N. A. (1967). *The Co-ordination and regulation of movements* (original work published 1935). Oxford: Pergamon Press.

Chollet, D. (1990). *Approche scientifique de la natation sportive. Bases biomecaniques, techniques et pschophysiologiques. Apprentissage, evaluation et corrections des techniques de nage.* Paris: Vigot.

Doignon, J. P. (1991, September). *Knowledge spaces and skill Assignments.* Paper read at the 22nd EMPG-meeting.

Doignon, J. P., & Falmagne, J. C. (1985). Spaces for the assessment of knowledge. *International Journal of Man-Machine Studies, 23*, 175–196.

Doignon, J.-P., & Falmagne, J.-C. (1998). *Knowledge spaces.* Berlin: Springer.

Falmagne, J. C., Koppen, M., Villano, M., Doignon, J. P., & Johannesen, L. (1990). Introduction to knowledge spaces: How to build, test and search them. *Psychological Review, 97*, 201–224.

Feltz, D. L., & Landers, D. M. (1983).] The effects of mental practice on motor skill learning and performance: A meta-analysis. *Journal of Sport Psychology, 5*, 25–57.

Gibson, J. J. (1979). *The ecological approach to visual perception.* Boston: Houghton & Mifflin.

Glencross, D. J. (1980). Levels and strategies of response organization. In G. E. Stelmach, & J. Requin (Eds.), *Tutorials in motor behavior* (pp. 551–566). Amsterdam: North Holland.

Hebb, D. O. (1949). *The organization of behavior.* New York Wiley.

Heuer, H. (1983). *Bewegungslernen.* Stuttgart: Kohlhammer.

Heuer, H. (1985). Wie wirkt mentale Uebung? *Psychologische Rundschau, Band XXXVI, Heft 3*, 191–200.

Heuer, H. (1990). Psychomotorik. In H. Spada (Hrsg.), *Lehrbuch allgemeine Psychologie* (pp. 495–559). Bern: Huber.

Keele, S. W. (1968). Movement control in skilled motor performance. *Psychological Bulletin, 70*, 387–403.

Kelso, J. A .S., Tuller, B., Vatikiotis-Bateson, E., & Fowler, C. A. (1984). Functionally specific articulatory cooperation following jaw perturbations during speech: Evidence for coordinative structures. *Journal of Experimental Psychology: Human Perception and Performance, 10*, 812–832.

Lashley, K. S. (1938). The accuracy of movement in the absence of excitation from the moving organ. *American Journal of Physiology, 43*, 169–194.

Lukas, J. (1997). *Modellierung von Fehlkonzepten in einer algebraischen Wissensstruktur.* [Modeling missonceptions in an algebraic knowledge structure]. *Kognitionswissenschaft, 6*, 196–204.

Maglischo, E. (1982). *Swimming faster. A comprehensive guide to the science of swimming.* Palo Alto: Mayfield Publishing Company.

Marshall, S. P. (1981). Sequential item selection: Optimal and heuristic policies. *Journal of Mathematical Psychology, 23*, 134–152.

Meijer, O. G., & Roth, K. (Eds.). (1988). *Complex movement behaviour: The motor action controversy.* Amsterdam: Elsevier.

Munhall, K. G., & Kelso, J. A .S. (1985). Phase dependent sensitivity to perturbation reveals the nature of speech coordinative structures. *Journal of the Acoustical Society of America, 78 (1)*, 38.

Narciss, S. (1993). *Empirische Untersuchungen zur kognitiven Repraesentation bewegungsstruktureller Merkmale.—Ein wissenspsychologischer Ansatz zur theoretischen Fundierung des Mentalen Trainings.* Unpublished doctoral dissertation, University of Heidelberg, Germany.

Reed, E.S. (1982). An outline of a theory of action systems. *Journal of Motor Behavior, 14*, 98–134.

Reischle, K. (1988). *Biomechanik des Schwimmens.* Bockenem: Fahrenham.

Reischle, K., & Spikermann, M. (1992). Biomechanical Analysis of Swimming Techniques. In O. Grupe, H. Haag, & A. C. Kirsch (Eds.), *Anthology of Sport Sciences in the Federal Republic of Germany*, (pp. 163–190). Champaign, Ill.: Human Kinetics.

Schmidt, R. A. (1975). A schema theory of discrete motor skill learning. *Psychological Review, 82*, 225–260.

Scott, D. (1982). Domains for denotational semantics. In M. Nielsen, & E. M. Schmidt (Eds.), *Automata, Languages and Programming*, (pp. 577–613). Heidelberg: Springer.

Stelmach, G. E., & Diggles, V. A. (1982). Control theories in motor behavior. *Acta*

Psychologica, 50, 83–105.

Turvey, M.T. (1977). Preliminaries to a theory of action with reference to vision. In R. Shaw, & J. Bransford (Eds.), *Perceiving, acting and knowing: Towards an ecological psychology*, (pp. 211–267). Hillsdale, NJ: Laurence Erlbaum Associates.

Turvey, M. T. (1994) From Borelli (1680) and Bell (1826) to the dynamics of action and perception. *Journal of Sport & Exercise Psychology, 16*, 128–157.

Unnewehr, J. (1992). Prozeduren der Wissensdiagnose (Benutzerhandbuch). Bericht aus dem Psychologischen Institut der Universität Heidelberg, Nr. 74.

Zimmer, A. C., & Koerndle, H. (1988). A model of hierarchically ordered schemata in the control of skilled motor action. *Gestalt Theory, 10*(1), 85–102.

Zimmer, A. C., & Koerndle, H. (1994). A Gestalt theoretic account for the coordination of perception and action in motor learning. *Philosophical Psychology, 7* (2), 249–265.

AUTHOR INDEX

Numbers in *italics* indicate pages with complete bibliographic information

A

Adams, J. A., 199, *218*
Airasian, P. W., 7, *11*, 42, *65*
Albert, D., 4, *12*, 32, 38, *39*, 42, 44, 54, 65, *65*, 73, 76, *99*, 106, *131*, 181, 182, 193, *194*, *195*

B

Bart, W. M., 7, *11*, 42, *65*, 68, *99*
Bernstein, N. A., 198, *218*
Birkhoff, G., 6, *11*, 18, 24, *39*, 76, *99*
Brainerd, B., 84, 88, *99*
Brown, J. S., 180, *194*, *196*
Burton, R. R., 180, *194*

C

Chéron, A., 29, *39*
Chollet, D., 206, *218*
Chomsky, N., 69, *99*
Cosyn, E., ix, *ix*, x

D

Düntsch, I., 105, *131*
Davey, B.A., 24, *39*
Degreef, E., 7, *12*, 68, *101*
de Groot, A. D., 26, *39*, 65, *66*
Diggles, V. A., 198, *219*
Doignon, J.-P., ix, *ix*, 4, 6–8, 10, *11*, *12*, 17, 18, 37, *39*, 42, 57, 65, *66*, 79, 81–83, 98, *99*, *100*, 104, 105, *131*, 134, *153*, 158, *178*, 181, 184, *194*, *195*, 198, 202, 203, *218*
Dowling, C. E., ix, *ix*, 8, *11*, 42, *66*, 68, *100*, 105, *131*

E

Edlinger, B., 35

F

Falmagne, J.-C., 4, 6–8, 10, *11*, 17, 18, 37, *39*, 57, 65, *66*, 68, 79, 81–83, 98, *100*, 104, 105, *131*, 134, *153*, 158, *178*, 181, 184, *195*, 198, 202, 203, *218*
Falstein, L. D., 118, *131*
Feltz, D. L., 215, *218*
Flament, C., 7, *11*
Fowler, C. A., 199, *219*
Fries, S., 7, *11*

G

Gediga, G., 105, *131*
Geisdorf, H., 29, *39*
Gibson, J. J., 199, *219*
Glencross, D. J., 199, *219*
Graesser, A. C., 94, *100*
Greene, E., 86, *101*
Guttman, L. A., 6, *11*, *39*, 73, *100*

H

Hebb, D. O., 199, *219*
Heines, J. M., 180, *195*
Held, T., 7, *11*, 24, 32, 37, 38, *39*, 42, 44, 54, *65*, *66*, 68, 69, 73, 86, 88, 92, *99*, *100*, 106, *131*, 182, 193, *194*, *195*
Hellriegel, P., 35
Herbst, K., 73, *100*
Heuer, H., 201, *219*
Hierholz, B., 28
Hron, A., 180, *195*

J
Johannesen, L., *11*, 17, 18, *39*, 81, *100*, 104, 105, *131*, 158, *178*, *218*
Judowitsch, M. M., 29, *40*

K
Kadijk, M., 65
Kambouri, M., *11*, 68, *100*
Keele, S. W., 199, *219*
Kelso, J. A. S., 199, *219*
Kimball, R., 180, *195*
Kintsch, W., 9, *11*, 84, 86–88, *100*
Klahr, K., 135, 136, *153*
Klein, D., 93
Koerndle, H., 198, 199, 208, *220*
Kondo, H., 180, *195*
Koppen, M., ix, *ix*, 8, *11*, *12*, 17, 18, *39*, 42, 65, *66*, 68, 81, *100*, 104, 105, *131*, 158, 174, *178*, 203, *218*
Korossy, K., 22, 37, 38, *39*, *40*, 42, *66*, 106, 110, 113, 114, 116, 118, *131*, 160, *178*, 181–183, 193, *195*
Kotovsky, K., 9, *12*, 133–136, *153*
Krause, B., 32, *40*
Krus, D. J., 42, *65*, 68, *99*
Kunichika, H., 180, *195*

L
Landers, D. M., 215, *218*
Lashley, K. S., 199, *219*
Lipski, J. M., 88, *100*
Lukas, J., 4, 8, *12*, 38, *40*, 65, 68, *100*, 106, *131*, 181, 182, *195*, 203, 218, *219*

M
Maglischo, E., 206, *219*
Maiselis, I. L., 29, *40*
Mandl, H., 180, *195*
Marcus, S., 88, *100*
Mariacher, H., 65
Marshall, S. P., 202, *219*
Matsuda, N., 180, *195*
Meijer, O. G., 199, *219*
Meserve, B. E., 118, *131*
Micka, R., *40*
Ministerium für Kultus und Sport, Baden-Württemberg, 128, *131*
Mörtel, P., 65
Müller, C. E., ix, *ix*, 8, *12*
Müller, H. R., 8, *12*
Munhall, K. G., 199, *219*
Musch, J., 35

N
Narciss, S., 200, *219*

O
O'Shea, T., 180, *195*
Okamoto, T., 180, *195*, *196*
Otsuki, S., 180, *195*, *196*

P
Ploetzner, R., 180, *195*
Priestley, H.A., 24, *39*
Ptucha, J., 35

R
Reed, E. S., 199, *219*
Regenwetter, M., 8, *12*
Reischle, K., 206, 214, *219*
Riha, J. R., 94, *100*
Roth, K., 199, *219*

S
Scandura, J. M., 73, *100*
Scheiblechner, H., 73, *101*
Schmid, A., 118, *132*
Schmidt, R. A., 199, *219*
Schneider, W., 94, *101*, 147, *153*
Schrepp, M., 7, *11*, 32, 38, *39*, 42, 44, 54, 63, *65*, *66*, 106, *132*, 134, 138, 145, 148, *153*, 182, 193, *194*, *195*
Scott, D., 203, *219*

Setek, W. M., 118, *132*
Simon, H. A., 9, *12*, 133–136, *153*
Sleeman, D., 180, *196*
Sobel, M. A., 118, *131*
Spada, H., 9, *12*, 180, *195*
Speckmann, W., 29, *40*
Spikermann, M., 214, *219*
Spreitzhofer, D., 35
Stelmach, G. E., 198, *219*

T

Takeuchi, A., 180, *195, 196*
Thiéry, N., ix, *ix*, x
Tuller, B., 199, *219*
Turner, A., 86, *101*
Turvey, M. T., 199, 200, *220*

U

Unnewehr, J., 210, *220*

V

van Dijk, T. A., 9, *11*, 84, 86–88, *100*
van Buggenhaut, J., 7, *12*, 68, *101*
van Leeuwe, J. F. J., 7, *12*, 42, 62, *66*, 68, *101*
Vatikiotis-Bateson, E., 199, *219*
Villano, M., 8, *11, 12*, 17, 18, *39*, 68, 81, *100, 101*, 104, 105, *131*, 158, *178*, 203, *218*

W

Wallace, J. G., 135, 136, *153*
Watanabe, K., 180, *195*
Wölk, M., 35

Y

Yasuda, K., 180, *196*

Z

Zimmer, A. C., 198, 199, 208, *220*

SUBJECT INDEX

A

Algebra problems, *see* Applications
AND/OR Graph, 79, 80
Antichain, *see* Order structures
Applications, 9–11
  algebra problems, 19, 22
  chess problems, 26–32, 51–54
  elementary arithmetic, 118–126
  elementary calculus, 5
  elementary stochastics, 84
  letter series completion, 133–152
  motor learning, 197–218
  number-series completion, 15, 32–37
  word problems, 67–99
Attribute clauses, 78, 79, 83
Attribute mapping, 78–82
Attribute order, 76, 81, 82, 91

B

Birkhoff's theorem, 6

C

Cartesian product, 23, 75
Chess problems, 26, 41–65, *see also* Applications
  list of, 29, 52
  motives, 26, 28, 42, 43, 51
  psychology of, 26
  three move problems, 26
Cognitive demands, 19
Coherence graph, 86, 87
Coherence, textual, 84, 85, 87, 92, 99
Cohesion graph, 88
Cohesion, textual, 84, 85, 87, 92, 99
Competence, 103–130
  elementary, 107
Competence partition, 162, 164
Competence structure, 106–109, 160
  representation problem, 112
Competence-based problem set, 108, 109
Competence-performance structure, 103–130, 156–177
Component, *see* Problem components
Component structure, 21, 22, 30
Component-based knowledge spaces, 15–38, 41–65
  principles for, 44
Componentwise distance, 80, 81
Componentwise product order, 76
Componentwise span, 80, 81
Congruence partition, 164–168
Congruence relation, 164–165
Correlational agreement coefficient, 62–64

D

Demand assignment, 82, 83, 91, 93, 98
Demand multiassignment, 83, 97
Diagnostic, 108–111, 160
  union-stable, 111–118, 159–162, 164, 169–173
Distance, 31
  between sets, 31, 55, 95, 97
  frequencies, 57, 60, 61, 95, 144, 149
  mean, 31, 36, 63
  minimal, 31, 36, 63
  symmetric, 145, 149
Distance agreement coefficient, 63, 64
Dominance rule, 24

E

Empirical test, 54–62, 93–97, 127–128, 142–152, 211–217
Equivalence relation, 162

F

Formal grammar, 69
Fundamental theorem for union-stable diagnostics, 168–173
Fuzzy competence assessment, 162–164

G

Generative grammar, 69
Goal–directed knowledge instruction, 159
Goodness of fit, 31
Guttman scale, 6, 16

H

Hasse diagram, 5, 51, 79, 80, 144

I

Inductive reasoning, 15–38
Information system, 203
Intelligent tutorial system, 179–194
  diagnostic component, 182, 184
  dynamical component, 181
  effectivity, 189
  statical component, 181
  student model, 182
  teaching component, 183
  teaching process, 184, 186
Interpretation function, 109, 125, 160
Item tree analysis, 7

K

Knowledge assessment, 16
Knowledge spaces, 17, 18, 105
  basis of, 104
  determination of, 7–9, 42
  examples of, 5, 30, 34, 35, 51–54
  introduction to, 3–11
Knowledge state, 18, 30, 104
Knowledge structure, 3, 4, 18, 104
  sensorimotor, 205

L

Learning, 156–177
Learning path, 174–176
Learning step, 161
Letter series completion, *see also* Applications
  difficulty of, 138–142
  solution process, 136–138
Level of recursion, 32
Linear order, *see* Order structures

M

Motor action, complex, 203–206
  breast stroke, 206–211
  formal description of, 204
Motor control and learning, 198–206, *see also* Applications
  assessment of, 201–206
Multiset, 46, 47

N

Number-series completion, *see also* Applications
  ambiguity, 36

O

Order filter, 113, 115
Order of problem presentation, 32
Order structures, 4, 161
  antichain, 16, 17, 19, 28
  Hasse diagram, 16, 17
  lexicographic order, 24, 25, 36
  linear order, 16, 20
  partial order, 16, 47, 167–168
  quasi order, 5, 16, 17, 45, 47, 49, 141
  semilattice, 165–167
Ordering principles, 57
  multiset inclusion, 42, 46–47, 59

sequence inclusion, 42, 47–49, 59
set inclusion, 42, 44–45, 59

P

Partial order, *see* Order structures
Performance, 103–130
Performance space, 108, 125, 126, 160
Performance state, 107–109, 160
competence-based, 109
Performance structure, 108, 110, 160
sensorimotor, 204
Phrase structure grammar, 69
Problem attribute, 73, 74
Problem coding, 69–76, 98
Problem components, 19–21, 28, 33, 73, 74, 83, *see also* Problem construction
attributes of, 22, 33
Problem construction, 15, 21, 69
combination of motives, 28
componentwise ordering, 15, 23, 33
principles of, 23, 70
sequence of motives, 48
set inclusion, 15, 19, 21, 28
varying attributes, 22
Problem demands, 81–83
Problem grammar, 72–74, 83, 88
Problem ordering, 17, 138–142, *see also* Problem construction
Problem solving, 15–38, *see also* Applications
Problem space, 44
Problem structure, 17, 21–23, 34, 76–83
Problems
equivalence, 110
union-stable, 113–115
Propositional text bases, 85–87

Q

Quasi order, *see* Order structures
Quasi-ordinal knowledge space, 18
Query procedures, 8

R

Representation function, 109, 125, 160–162
Response patterns, 57, 59–61, 95, 96, 145, *see also* Solution patterns

S

Skill assignment, 81, 82, 179–194
Solution patterns, 105, 107, *see also* Response patterns
admissible, 6
Solution times, 32
Student modeling, 158
Surmise function, 104, 126, 182
example of, 122
Surmise mapping, 79
Surmise relation, 7, 18, 21, 23, 28, 37, 45, 47, 76–78, 140, 202
example of, 46, 48, 50, 144, 149
Surmise system, 7, 78–81, 104, 105, 202

T

Text comprehension, 84
Text production, 84

U

Union-preserving representation, 112, 113

V

Violational coefficient, 63, 64